Dedicated to the memory of my grandfather,

Dr. Kenneth T. Bird

BENEATH THE NORTH ATLANTIC

by Jonathan Bird

Tide-mark Press
Hartford, Connecticut

   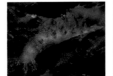   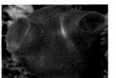   

Published by Tide-mark Press Ltd., P.O. Box 280311, East Hartford, CT 06128-0311
Distributed in Canada by Monarch Books of Canada
Distributed in the U.K. by Lavis Marketing
Design and typography by Jane Kirk
Printed in Singapore by Craft Print

First Printing

ISBN 1-55949-314-3 Library of Congress Catalog Number 96-60420

CONTENTS

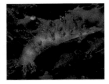

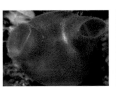

PHOTO CREDITS

A LITTLE LATIN FOR TAXONOMY

Speakers of English use Latin with surprising frequency. The roots of many English words come from Latin, as do many words and phrases which we use virtually unchanged. In zoology, since the taxonomic nomenclature used throughout the world is based on Latin, it is useful to know the correct pluralization and pronunciation. The following guides may be helpful in coping with Latin.

Words which end in "a," like "bacteria," are pronounced "ah." Words which end in "ae" are pronounced "eye" (example: larvae is "lar-veye," not "lar-vee") Words which end in "i" are pronounced "ee" (example: alumni is "al-um-nee," not "al-um-neye"). .

What about singular and plural uses? Latin words can be masculine, feminine or neuter. Masculine words generally end in "us," like the word "alumnus." The plural of masculine words adds "i," or "alumni." So if someone says "I am an alumni of this college," he must have a split personality!

Feminine singular words usually end in "a" (like "larva") and the plural ends in "ae" (larvae). Neuter singular words generally ends in "um" (example: bacterium), and the plurals ends in "a" (bacteria). The neuter plural form is especially confusing because it looks just like the feminine singular. The only way around that problem is to know the gender of the words in question.

"foot" have roots like "pod" or "ped," while the root from Greek is "pus"). What is the correct plural? Many say "octopi," since the singular looks like a Latin masculine singular noun which ends in "us" It really isn't, so pluralizing the word "octopi" is not correct. (Most people also mispronounce it. If it <u>were</u> "octopi," it would be pronounced "octo-pee" not "octo-pie.") The correct

	Singular	**Plural**
Masculine	-us	-i
Example	nauplius	nauplii
Feminine	-a	-ae
Example	amoeba	amoebae
Neuter	-um	-a
Example	flagellum	flagella

Finally, there are words used in English which do not come from Latin, but look like they do, so they are often erroneously "Latinized." Octopus is the most common example. The word actually comes from Greek, not Latin (in English words from Latin related to

plural is "octopuses," believe it or not. I prefer not to use either one, since every time you say "octopuses" someone will disagree. Instead, I use "octopods," which is the proper plural of the (Latin) taxonomic order Octopoda, containing the eight-armed cephalopods.

ACKNOWLEDGMENTS

A great many people have helped me with this book along the way, and I offer my most sincere thanks. Among the people to whom I am grateful are: my dad, first for getting me interested in diving, and second for giving me my first underwater camera; my mom; Roger Allen, Ken McGagh, Jim Matulis, David and Susan Millhouser, and Brian Skerry for their substantial photographic contributions (without which this book would be a lot thinner); Scott Kaeser and the staff of Tide-Mark Press for giving me the opportunity to undertake this project and prove that there is something special in the North Atlantic; Dr. Larry G. Harris, professor of zoology at the University of New Hampshire, for his editing of the text, insightful tidbits of information, and assistance in identifying some of the photographs; Jean Bird; Jack and Elsie Lambert; Jeffrey Bird; the staff of Oceanic Research Group (ORG; especially Tom Krasuski, who trekked to the far reaches of the North Atlantic with me); the helpful people at Cape Ann Divers in Gloucester, Massachusetts, especially Matt Kripp, Mary Blake, and Dave Stillman; my North Atlantic dive buddies past and present (Rick Doyle, Phil Kelley, Liz Osborne, Alan Restell, Eric Rosenkrantz, Jay Solomon, and many others); Rebecca Farr of the NASA Goddard Space Flight Center, Distributed Active Archive Center, for helping me obtain satellite images; Dr. Kenneth Baldwin, Dr. Janet Campbell; Scott Kraus of the New England Aquarium; Bob Evans of Force Fin; my friends at the Receiver/ Exciter Department at Raytheon; all the schools that have sponsored my lectures; John Prenier and Jim Prenier at East Coast Fish Farms; Cecil Cates; Owen Lawlor; Mary Pottle; Chief Thomas Dutton, of the United States Coast Guard for allowing ORG to hitch a ride to Mount Desert Rock; Bob Bowman and Dr. Steven Katona of the College of the Atlantic for their help with seals; Tim Cole, Lisa Cuellar, Stephanie Martin, Allen Reitsch, and Ann Zoidis of the Mount Desert Rock Whale and Sea Bird Research Station; Chris and Elaine Eaton at Harbor Divers in Bass Harbor, Maine; Stefan Arnold; Art Cohen, Cheryl Evans; Vito Giacalone of the Massachusetts Department of Marine Fisheries; Heather Hoffman; Stephen Nathanson; Bob Stapel; Kathy Sterling; Tracy Sundell at the Nantucket Marine Department; Cory LaJoie at Northeast Dive Journal; the Boston Sea Rovers; Captain Jacques-Yves Cousteau for his inspiration; and finally, my wife, Kimberly, who put up with (and encouraged) me throughout this project.

ABOUT THE AUTHOR

Jonathan Bird is the president of Oceanic Research Group, a nonprofit environmental organization dedicated to the conservation of the world's oceans and marine life through education. He has directed many educational films, including *Beneath the North Atlantic, Beneath the South Pacific, Seals of New England, The Gentle West Indian Manatee,* The *Coral Reef: A Living Wonder,* and *Sharks: Masters of the Seas.* Jonathan has designed and hosted several educational interactive television programs for the Massachusetts Corporation for Educational Telecommunications (MCET), including the hit series *The World Beneath the Sea.* He is currently completing a masters degree in ocean engineering at the University of New Hampshire and is working on new films with Oceanic Research Group. He has lectured extensively on marine subjects in schools, museums, dive conventions, and aquaria.

Oceanic Research Group educational films are designed for use in schools, and are available through AIMS Media, 9710 DeSoto Avenue, Chatsworth, California, 91311-4409, telephone (800) 367-2467.

GLOSSARY

aboral - On a sea star, the surface of the animal opposite the mouth, usually the top or dorsal side.

amebocyte - In a sponge, the type of cell responsible for transport of food within the animal.

anadramous (anad´ r əm əs) - A type of fish which is born in fresh water but then migrates to the ocean to live.

anthozoan - Meaning "flower-animal", this term applies to any member of the cnidarian class Anthozoa, including anemones and corals.

arthropod - A phylum of animals characterized by a jointed exoskeleton made of chitin including crabs, copepods, lobsters, shrimps, etc.

autotomy - In sea stars, meaning the loss of a limb.

baleen - A material used by the filter-feeding whales as a filter to catch plankton and small fishes. Baleen is made of keratin, the same material from which human fingernails are made.

benthic - Pertaining to the sea floor.

benthos - Flora or fauna living on or in the ocean bottom.

biomass - Meaning the total mass of a biological population.

brachiopod - An animal belonging to the ancient phylum Brachiopoda. Brachiopods are bivalve animals outwardly resembling clams or mussels, but have completely different internal anatomies.

cephalothorax - Meaning "head-body," the part of an arthropod which contains the head organs (eyes, mouth etc.) and body organs.

cheliped - The scientific term for the claw of an arthropod.

chitin - Material of which mollusk shells and arthropod exoskeletons are made.

choanocyte - A collar cell of a sponge. Each collar cell has a flagellum used to move water through the sponge and a funnel-shaped collar to trap food particles.

cnidarian (nid air ee an) - Any member of the phylum Cnidaria. All cnidarians have stinging cells called nematocysts.

coelenterates - An antiquated term for cnidarians no longer in widespread use because it includes the ctenophores which are now classified into their own phylum.

copepod - A small planktonic arthropod with a simple body, single eye, and a pair of long antennae used for locomotion. The copepod is the single most numerous animal on Earth, and is important food for many fishes and whales. It would take about 200,000 copepods to fill a coffee-cup.

crustacean - Any animal belonging to the arthropod's subphylum Crustacea, including crabs, lobsters, shrimp, etc.

ctenophore (ten´ə for) - Any animal belonging to the phylum Ctenophora. These are jellyfish-like animals which move by means of bands of cilia along their bodies. Although they resemble the scyphozoans, they lack nematocysts, and thus are not cnidarians.

echinoderm (i kī´ nə durm) - Any animal belonging to the phylum Echinodermata, meaning "spiny-skinned" and including the sea stars, sea cucumbers, sea urchins, and feather stars.

echolocation - A means of detecting distant underwater objects using sound. Animals which can echolocate (like dolphins) produce sounds (usually clicks) which travel through the water and bounce off of objects or prey. The echoes reflected back are interpreted and give the sender a mental picture of what lies ahead. Many animals can hunt and navigate exclusively using echolocation.

epifauna - Benthic fauna living on top of the substrate.

euphotic zone - The surface waters of an ocean region which receive enough sunlight to support photosynthesis (on average, to a depth of about 100-200 metres).

fission - In sea stars, when a sea star tears itself roughly in half. This is a means of reproduction.

fissiparous - A term describing sea stars which reproduce using fission.

hemocyanin - A copper-based pigment found in arthropods which causes the blood to appear blue.

hemoglobin - An iron-based pigment found in mammal blood which causes the blood to appear red.

hermaphroditic - Having both male and female reproductive systems in the same individual.

holoplankton - Any type of planktonic organism which remains planktonic for its entire life.

hydromedusa - A jellyfish stage in the life cycle of a hydrozoan.

hydrozoan - Any animal belonging to the cnidarian class Hydrozoa.

infauna - Animals that live within bottom sediments.

krill - A generic term used to describe any of a number of small shrimp-like plankton

macroplankton - Plankton which is physically large enough to be seen with the naked eye.

madreporite - The interface between the water vascular system of an echinoderm and the ocean.

medusa - The body form of a jellyfish.

medusoid - A term used to describe something which is medusa-like in appearance.

meroplankton - Any type of planktonic organism which is only planktonic for part of its life cycle.

nauplius - An early stage of development in arthropods such as copepods.

neap tide - The tide with the smallest tidal range for a particular area.

nektobenthic - A term applied to creatures which live on the ocean bottom but can leave it if they want, such as flounders.

nekton - Any swimming animal which lives up off the bottom but does not float at the surface.

neritic province - The ocean region from the shore line to the edge of the continental shelf.

neuston - Animals which float at the surface (such as the Portuguese Man-o-war).

notochord - A stiff dorsal cord present in urochordates at some point in their lives which provides support like a backbone.

nudibranch (noo də bronk, or noo də brank) - A marine gastropod, in most cases similar to a snail without a shell.

ocellus - A primitive light sensing organ, such as in a jellyfish.

oscula - The excurrent hole in a sponge.

ostia - The incurrent pores in a sponge.

oviparous (oh vip´ ərəs) - A type of reproduction where the young develop in eggs laid by the parent.

ovoviviparous (ō´ vo vī vip´ ərəs) - A type of reproduction where the young develop in eggs held internally by the mother until they hatch.

pelagic - pertaining to the open ocean.

phytoplankton - Plant plankton

picoplankton - Any microscopic plankton, such as protozoans.

planktivorous - Referring to any creature which consumes plankton.

plankton - Organisms which live suspended in the water column but are unable to counter water currents because of small size or limited motility.

planktonic - Pertaining to plankton, plankton-like.

planula - An early stage of development in cnidarians.

polyp - One of two basic body styles in cnidarians, where the cylinder-like body is anchored to the substrate and the mouth and tentacles are pointing away from the substrate.

polypoid - Meaning polyp-like in shape or appearance.

scyphozoan - The class of cnidarians containing the jellyfishes (more properly called the jellies).

sessile - Meaning unmoving, usually describing an animal which lives in one place on the bottom, such as a sponge.

spicule - A needle-like splinter which serves as the basic skeletal unit in certain sponges and soft corals.

spring tide - The tide with the largest tidal range for a particular area.

statocyst - The gravity sensing organ of a jellyfish.

taxonomy - The study of classification of living things.

test - An encasing or shell-like skeleton.

thermocline - The boundary between layers of water at different temperatures.

urochordate - The subphylum of chordates containing the sea squirts, tunicates and salps.

vibrissae - The whiskers of seals and sea lions which are sensitive to vibration.

viviparous (vī vip´ ərəs) - A type of reproduction where the young develop in a uterus and are born live.

water column - The general term for the location anywhere in the water between the surface and the bottom, but not floating on the surface or sitting on the bottom.

zooplankton (zōə plank tən) - Animal plankton.

WHAT IS SO SPECIAL ABOUT THE NORTH ATLANTIC?

Many people think of the North Atlantic Ocean as merely a rich fishing or dumping ground for the waste of eastern North America. Most people have probably never looked beyond the ocean's shores and boating surface. Yet beneath the waves of the Atlantic lies a vast world, largely unseen and unknown. This book will give you only a small glimpse of the huge diversity of life that exists in the frigid waters of the North Atlantic.

We know that more 70 percent of the Earth's surface is covered by water. The oceans support some two hundred thousand species, including representatives of all living phyla known to science. Neither terrestrial nor freshwater habitats can claim such diversity.

Ocean temperatures range from around 28° to about 90°F, a much narrower temperature variation than is found on land, and the oceans offer a relatively stable environment compared with other habitats. Despite its being relatively stable, however, the ocean is not the same throughout. Anyone familiar with the ocean around New England and who has visited Florida's Atlantic shore will probably recognize the considerable difference in both the appearance and temperature between them. The clear, warm waters of tropical seas draw so many vacationers and photographers that many people do not think of the temperate and polar seas as places filled with life, but they are. Consider for a moment where some of the most productive fishing grounds in the world are located:

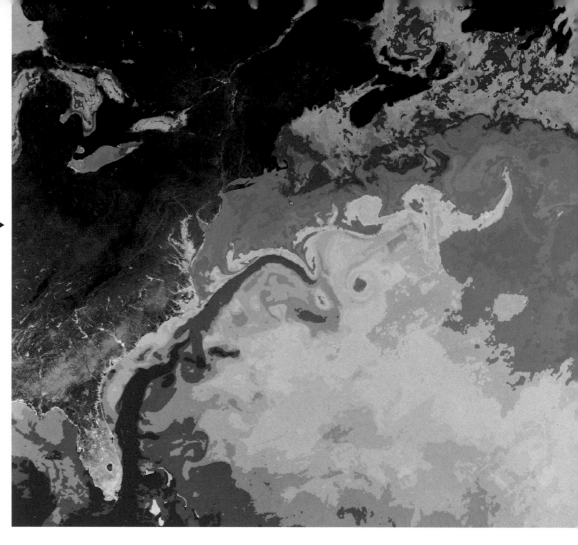

A false color image ▶ of the North Atlantic sea surface temperature as seen from a satellite. Warm water appears red to orange, while cold water is blue to purple. The Gulf Stream is easily seen here, as is the cooler water of the Labrador Current.

Argentina, Iceland, New England, Norway, Peru, and other cold-water regions.

Cold water and high latitudes promise dense fish populations because cold water (usually found in higher latitudes) tends to be more fertile than warm water. The fertility of an ocean region determines how much life it can support, and the entire system depends on the food chain; the lowest organisms in the food chain make the subsistence of the higher organisms possible. In the oceans, plankton are the lowest organisms in the food chain. Fertile oceans have large quantities of plankton, which support a massive web of higher life forms. Productive fishing grounds are characterized by the presence of large quantities of higher life forms, like pelagic fish.

The word *plankton* comes from a Greek word meaning "to drift." The definition of a plankter, therefore, is any organism that drifts with the sea. Contrary to popular belief, plankton are not necessarily microscopic; in fact, quite a few are not.

There are two main types of plankton: plant plankton, or phytoplankton, and animal plankton, or zooplankton. Phytoplankton are among the most important organisms in the oceans. There are thousands of different types of phyto-plankton, but the most common is the simple diatom. Although it reaches only about 80 micrometers [1 micrometer = one millionth of a meter], the diatom is food for thousands of other organisms, most of which are zooplankton. One estimate suggests that the phytoplankton of the world's oceans are responsible for producing 60 percent of the oxygen we breathe.

Zooplankton can take many forms: radiolarians, shrimplike copepods, arrow worms, dinoflagellates, the larvae of fish and other larger species, jellyfish, and so on. There are some carnivorous zooplankton, but most are herbivorous and eat phytoplankton. The existence of both carnivorous and herbivorous zooplankton, however, is entirely dependent on a good supply of phytoplankton.

Just about every other creature in the ocean is somehow involved with plankton. Even the largest creatures in the seas, the baleen whales, eat zooplankton by straining it from the water. Many whales eat more than a ton of plankton each day. Other creatures are dependent on the zooplankton indirectly because they eat something that lives on plankton. The great fishing grounds of the world are as highly productive as they are simply because they have a large supply of plankton to support life. But what circumstances make possible the growth of the plankton?

At first it might seem that plankton would be abundant in the tropical seas, owing to the warm climate and ample sunlight. After all, phytoplankton are plants and need sunlight. Sunlight, however, is only part of the equation. Gardeners know that plants need more than just water and sun. For garden plants to grow, they need decent soil. Soil provides the garden plant with a method of anchoring itself and a source of nutrients. Phytoplankton do not need to anchor themselves, but they do need nutrients. Both garden plants and plankton need the primary plant nutrients of nitrogen, phosphorous, and silicon compounds.

Nitrates, phosphates, and silicates originate in rocks or sediments that are eroded by the action of the sea. As they become available, these nutrients are quickly consumed by phytoplankton and other sea plants, such as the kelps and seaweeds. Nutrients are in short supply in the oceans because the supply is limited and because they are all contained within the cells of living creatures. Most of the nutrients needed by phytoplankton come from two sources: the decomposing feces of fish and zooplankton and the decomposed creatures themselves. The steady raining down of decomposing matter in the ocean has been termed *marine snow* because it looks much like falling snowflakes.

Marine snow sinks toward the sea floor, releasing nutrients along the way. Unfortunately, decomposing organic matter tends to sink faster than it can decompose, and it usually releases most of its nutrients at depths far below the euphotic zone, which is the upper 300 feet (100 meters) of water, where light penetrates sufficiently to support photosynthesis. If nutrients are stuck at the bottom of the sea and unavailable to phytoplankton, the plankton cannot grow, even if there is plenty of light.

The tropical oceans have very little phytoplankton because an effect called thermoclines (boundaries created by layers of water at different temperatures) traps nutrients at depths far below the euphotic zone. Surface waters of the tropical oceans are warmed by the sun all through the year and become much warmer than deep water. The warm water tends to float on top of the denser cold water, much like oil floating on vinegar in a salad dressing, creating thermoclines. The warm surface water cannot mix with the nutrient-rich deep water, and nutrients remain unavailable to phytoplankton. Thus thermoclines are of great significance to many creatures in the ocean.

In the temperate and polar seas, the winter is so cold that the surface water is at about the same temperature as the deep water. There are no significant thermoclines. This allows thorough mixing to occur between the layers, and nutrients are made available to the euphotic zone. Unfortunately, the polar seas have no light at all in their respective winter seasons, and the temperate seas have very short days during this time. In winter in these regions, phytoplankton production is low because of a limited amount of sunlight.

As spring gets under way and the days grow longer, the polar and temperate seas experience what is known as a spring plankton bloom. It is during this time that phytoplankton, with the benefit of both sunlight and nutrients, explodes into bloom. As the phytoplankton population grows, so does the population of everything that feeds on it (mostly zooplankton). Many creatures time their spawning cycles to coincide with the plankton blooms so that their offspring will have enough food to survive.

Because phytoplankton are plants and utilize chlorophyll, they are usually green in color. The huge quantities of photosynthetic phytoplankton suspended in the water give the fertile oceans their green tint. Tropical oceans appear blue because they have almost no particles suspended in the water. The water is quite pure, and pure water is least able to absorb blue wavelengths. It is easy to judge the fertility of an ocean region merely by examining its color.

During the past decade, scientists have used satellites to monitor surface ocean biological productivity, using special cameras to measure the color of the ocean water. This technique revealed that the North Atlantic has a spring bloom larger than in any other ocean on Earth.

Another important characteristic that encourages the growth of both types of plankton is cold water's ability to hold more dissolved gasses than warm water can. A spring bloom will produce much phytoplankton, which needs carbon dioxide to survive, as does any plant. If the water cannot hold enough carbon dioxide for the phytoplankton (and oxygen for the zooplankton), the bloom will die off.

As summer approaches in the polar and temperate seas, the spring bloom begins to wane as the hot sun and longer days warm the surface water, creating a thermocline. Although this thermocline is not as pronounced as the one in the tropics, it is enough to limit the exchange of nutrients between water layers. When its nutrient supply is cut off, the plankton population cannot continue to expand.

In the fall, as the days get shorter and the surface water cools, the thermocline dissipates, and the nutrient supply is reestablished. Phytoplankton begins growing again during a period called the fall bloom. This bloom is always less robust and shorter-lived than the spring bloom because the hours of sunlight decline as winter advances.

The seasonal plankton cycle makes underwater visibility variable throughout the year in the polar and temperate seas. Under ideal conditions, visibility in tropical oceans is as much as 200 feet (61 m), but 50-100 feet (15-30 m) is common throughout the Caribbean and South Pacific. In New England, visibility is rarely better than 10-15 feet (3-4 m) in the summer. Although some harbors in the North Atlantic are quite polluted, New England's lack of water clarity is the result of plankton production, not pollution, as many people seem to believe. In the winter and early spring (when plankton production is low), the visibility in New England can be as much as 60 feet (18 m), although 30-35 feet (9-10 m) is more common. The Arctic, which has visibility similar to New England's in the summer, has the best visibility in the world during the winter and early spring, when visibility frequently reaches 300 feet (91 m). Unfortunately, the temperature of arctic water in early spring is only about 29°F; New England water at that time is just slightly warmer, ranging between 30° and 38°F. These temperatures make for chilly diving that requires specialized cold-water gear.

The oceans are always in motion, never still like ponds. Ocean currents constantly move seawater, sometimes bringing cold water into warmer regions and vice versa. Plankton moves with these ocean currents.

On the east coast of the United States there are two major currents. The Gulf Stream is perhaps more well known than the Labrador Current. The Gulf Stream brings warm water up the coast from the Caribbean. Contrary to popular belief, the Gulf Stream never actually reaches New England or the northwestern Atlantic. If it did, the water in the northeast United States would be much warmer than it is. In fact, the northeast coast of the United States is bathed in the cold water of the Labrador Current as it makes it way southward from the Arctic.

The Labrador Current meets the Gulf Stream near the coast of Virginia, where the two currents join and head out to sea (see page 1). The Labrador current thus deflects the Gulf Stream across the Atlantic, where it finally washes the beaches of Ireland and Britain. Therefore the water on the west coast of Britain is much warmer than it is in New England, and Britain is significantly farther north than New England.

Benjamin Franklin, while working as deputy postmaster for the colonies of New England, noticed that it took two weeks longer for a postal ship to sail from England to the colonies by a northerly route than it did to sail by a much longer southerly route. His nephew, Timothy Folger, who was a Nantucket whaling captain, knew from experience of the warm Gulf Stream waters to be found at sea and drew a map for Franklin. In 1769, this map was first printed as an aid to speed navigation across the Atlantic. Captains got their ships into the Gulf Stream to get to England and avoided it to get back to the colonies.

Occasionally small warm rings (called eddies) of the Gulf Stream break off and swirl into Connecticut, southern Massachusetts, New York, and Rhode Island. These eddies can be more than 100 miles (160 km) across. Frequently, along with warm Gulf Stream water, they bring

juvenile tropical fish from North Carolina and southward. Volunteers from the New England Aquarium collect these fish and take them to the aquarium to be raised, since they could not survive the drastic temperature change in the North Atlantic's water at the onset of winter.

Cold eddies can also form from little swirls of Labrador Current water, which wash into the Gulf Stream. The result is a pocket of nutrient-rich, green water which sits in the middle of the Sargasso Sea, a clear, blue area of the Atlantic east of the Gulf Stream. These cold eddies can take more than a year to dissipate and carry with them the marine life that thrives in cold water. Cape Cod is a natural barrier to these two ocean currents. The Labrador Current keeps the water north of the Cape cold, while Gulf Stream eddies can keep water south of the Cape warmer by as much as 10°-15°F.

The Labrador Current is extremely important to New England. When the New World was first discovered, the North Atlantic's rich potential as a fishing ground quickly became apparent. The cold Labrador Current keeps that part of the North Atlantic generously supplied with nutrient-rich and plankton-filled arctic water. Hence, marine life flourishes in New England's waters (although overfishing over the past hundred years has taken a severe toll on some important commercial fish).

The currents and the plankton cycle in the North Atlantic are the basis for an extremely large and diverse population of marine life. For centuries, anglers have known that New England waters are among the best in the world for food-fish production, yet little attention has been paid to New England's other marine life. In this book I bring to light many of the beautiful and interesting creatures that call the North Atlantic home. It is my hope that exploring the myriad species of biologically diverse creatures living in the North Atlantic will win greater respect and concern for the delicate balance of life in this sea. It is with that hope that I present *Beneath the North Atlantic*.

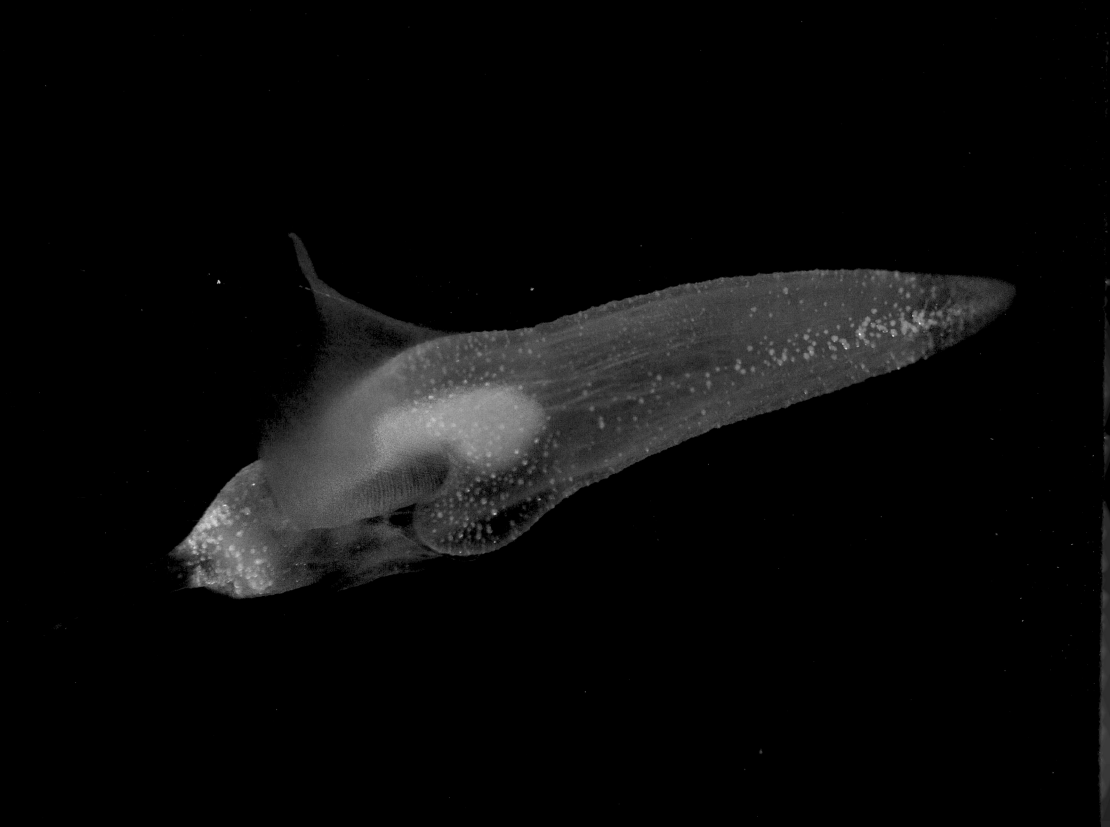

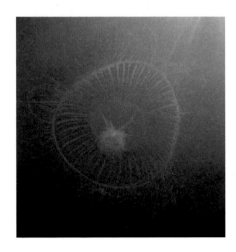

PLANKTON

The importance of plankton cannot be overemphasized. Although there are many more animal species on land than there are in the ocean, plankton (the word cromes from Greek and means to drift) include the largest biomass on Earth. Pound for pound, the total mass of zooplankton in the oceans exceeds the combined mass of all other animals on Earth. This huge volume of living matter is critical to the survival of every creature in the ocean. Realizing that phytoplankton represents such a large percentage of the plant life on the planet, we owe plankton more than a little respect.

Phytoplankton are certainly drifters; they are plants that move with the ocean currents and are food for zooplankton. Zooplankton are tougher to characterize because there are so many different types and because many of them actively swim. That would seem to contradict the very definition of plankton as drifters, making them *nekton*, or swimmers, from the Greek word 'to swim.'

For example, is the common jellyfish planktonic or nektonic? Well, it swims, so it must be nektonic. But because it is a weak swimmer, it also goes where the current takes it. Maybe it is planktonic after all. Most scientists recognize weak-swimming animals as plankton; that is, if it cannot swim against a current, it is planktonic. This revised definition of plankton includes many different types of creatures (including scuba divers).

A popular misconception is that all plankton are small. Size is relative. The zooplanktonic krill, which are shrimplike crustaceans, reach 1-2 inches (2-5 cm) in length. They dwarf the copepods, which are less than $1/4$ inch (5 mm) long, which in turn dwarf the radiolaria ($1/2$ mm), which in turn dwarf the tintinnid (50 μm). Jellyfish are planktonic, however, and one species has been recorded at more than 6 feet (2 m) across, with tentacles almost 200 feet (61 m) long. Therefore, plankton ranges in size from the microscopic to the gigantic.

◄ *The Sea Butterfly* (**Clione limacina**) *is a one inch mollusk found predominantly in the open ocean, where large numbers are eaten by whales.*

 # PLANKTON

 ## TYPES OF ZOOPLANKTON

It is safe to say that every phylum of ocean creatures has at least a few members that are planktonic at some point in their lives. For example, nearly all echinoderms begin life as planktonic larvae, floating in the water column until they grow large enough to settle down and mature into sea stars or sea cucumbers. (The water column is anywhere in the ocean above the bottom and below the surface.) Crustaceans have a *nauplius* stage of development (a larval form) that is similar. The planulae of hydroids and corals are planktonic. Many animals release eggs that are planktonic, and even the hatchlings are planktonic for a time. Most newborn fish (*fry*) are planktonic because of their extremely small size and limited swimming power. Sea squirts and tunicates have a *tadpole* stage that is planktonic. Even sponges have a short planktonic stage. All these creatures eventually become swimmers, but until they do, their risk of being eaten is high. Each one is part of a complicated food web, whereby only the most fit (or lucky) individuals live long enough to mature, reach their adult form, and reproduce. This plankton is called meroplankton.

Other animals remain planktonic for their entire lives. They are called holoplankton and come from many phyla. Although there are myriad species of microscopic, single-celled (*protozoan*) holoplankton, like bacteria, amoebae, radiolaria, and foraminiferae (called picoplankton), I limit my discussion here to some of the larger holoplankton,

called macroplankton.

The copepod is the single most abundant animal on Earth. Copepods are tiny crustaceans, only a fraction or an inch, or a few millimeters, long that are related to crabs and shrimp. They spend their entire lives swimming in the water column seeking tiny bits of phytoplankton for food and becoming food for many small fish and invertebrates. Although few people will ever see one, these animals are the most numerous of all zooplankton and are extremely important in the food chain. *Calanus finmarchicus* is the most abundant copepod in the North Atlantic and is a favorite food of whales. A Right whale may eat a thousand or more pounds of these copepods every day. Considering that *C. finmarchicus* is smaller than a grain of rice, this represents a lot of copepods!

Krill are another important zooplankton that occur in large schools. Although "krill" refers collectively to many kinds of planktonic crustaceans, they are actually more closely related to amphipods. They are another principal food of baleen whales, which swim though a school of krill and strain them from the water.

A particularly beautiful zooplankter is the Pteropod, commonly called the Sea Butterfly. One species of this mollusk (*Clione limacina*) is common in the North Atlantic. Resembling a tiny, purplish squid lacking tentacles, this 1-inch (2$^1/_2$ cm) gastropod swims gracefully through the water using two winglike fins on the front of its body. It is abundant offshore and is an important food for whales in the North Atlantic. The Sea Butterfly feeds primarily on

a more common and much smaller pteropod, which in turn survives on phytoplankton.

 ## SCYPHOZOA

The scyphozoans are the class of jellyfish in the phylum Cnidaria. Scyphozoa literally means 'bowl animal,' from their bowllike appearance. Although commonly called jellyfish, scientists sometimes prefer to reserve the term *fish* for true fish and call the jellyfish *jellies*. Since jellyfish are cnidarians, they are closely related to sea anemones and, like all cnidarians, have stinging nematocysts (see chapter 3). Although some jellyfish do have powerful stings, most are incapable of seriously injuring a human being. The sting of the jellyfish is powerful enough to capture food and fend off fish and other creatures that might want to make a meal of it. Even so, there are creatures in the ocean that make the jellyfish a prime source of food at least some of the time.

The Ocean Sunfish (*Mola mola*) can grow to 13 feet (4 m) in length and 3,300 pounds (1,500 kg) and feeds almost exclusively on jellyfish and ctenophores. Its tiny mouth is all that it needs to eat these animals. The Leatherback Turtle (*Dermochelys coriacea*), which is the only pelagic reptile in the North Atlantic, also eats jellyfish at least part of the time. Both of these two large pelagic animals journey to the North Atlantic in the summer because of the abundance of scyphozoans.

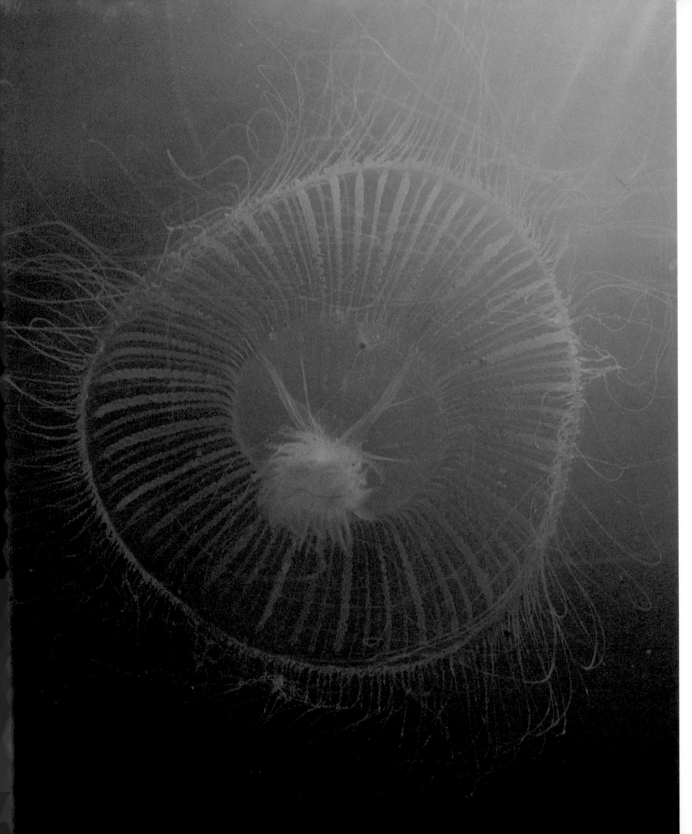

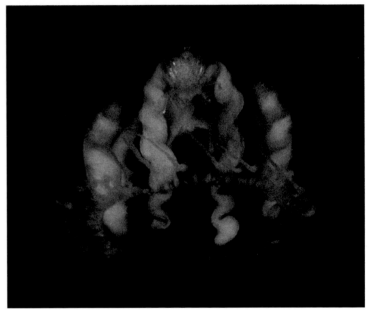

▲ The Eight-ribbed hydromedusa
(**Melicertum octocostatum**) grows to
only one inch across and is shed from its
hydroid colony to go forth and settle
elsewhere, thus creating a new colony.

◄ The Many-ribbed hydromedusa (**Aequorea
aequorea**) looks just like a scyphozoan
(jellyfish), but is technically a hydrozoan with
no known polyp stage. It is bioluminescent and
shimmers in the water column after dark.

PLANKTON

The most common jellyfish in the North Atlantic is probably the Moon Jelly (*Aurelia aurita*). Although they can reach a diameter of about 16 inches (40 cm), they are typically much smaller (6-8 inches, or 15-20 cm). This animal has four distinctive semicircular lobes within its body that serve as gonads. Around the circumference of the animal are many short tentacles containing nematocysts. Although not as toxic to humans as the *Physalia* (see chapter 3), contact with the Aurelia may cause a mild rash and itching. Observed in the wild, the *Aurelia* is a beautiful and graceful creature. It swims with a pulsating motion, contracting its bell (main body) to force water out, opening again, and closing. Although the Moon Jelly is carried at the whim of ocean currents, it does have enough mobility to adjust its depth and direction. Jellyfish have an organ called an *ocellus* to detect the presence and direction of light and an organ called a *statocyst* to detect the direction of gravitational pull. Using these two types of organs, the jellyfish always knows which way is up; when turned upside down a jellyfish can always right itself upright again.

The Sea Nettle (*Chrysaora quinquecirrha*) is another common scyphozoan in the Atlantic, from Cape Cod to Florida. These jellyfish can measure 10 inches (25 cm) across, although they are frequently smaller. Their sting will irritate human skin.

Although generally the medusa form of hydroids are called hydromedusae (see chapter 3), the Many-Ribbed Hydromedusa (*Aequorea aequorea*) has no known polypoid stage. We do know, however, that this animal is in fact a hydro-zoan, like the Portuguese Man-of-War, and not a true jellyfish (scyphozoan), like the Moon Jelly. Even so, this common worldwide species looks and acts like a scyphozoan and grows to about 7 inches (18 cm) across. It is found in the Atlantic from Maine to Florida and throughout the Caribbean and the Gulf of Mexico. Perhaps the most interesting feature of the animal is its bioluminescence, which means that it has the ability to create light chemically, much as a firefly does.

CTENOPHORA

In outward appearance, ctenophores are similar to jellyfish, but they lack the jellyfish's stinging nematocysts. These filter-feeders swim by means of eight ciliated bands, or rows, of cilia, along the outside of their bodies. There are about fifty species known to science worldwide. The rows of cilia look like little combs and suggest their common name, Comb Jellies. (In fact, *Ctenophore* is Latin for 'Comb-Jelly.')

Many species of ctenophores are bioluminescent. On dives in Rhode Island, I have been surrounded by thousands of Leidy's Comb Jellies (*Mnemiopsis leidyi*), which are bioluminescent. They will pulse with light upon contact with something. At night I can turn off my dive light and swim through the pitch black water while glowing bulbs of light bounce off of my body. It is eerie, but spectacular to observe.

Bioluminescence is not reserved for ctenophores, of course. Many tiny plankton are bioluminescent and will light up as a divers' fins and hands sweep past. This bioluminescence is extremely dim and is discernible only to the naked eye under very dark conditions. It has proved impossible for me to photograph it.

There are many species of ctenophores in the North Atlantic, but easily the most common is the Northern Comb Jelly (*Bolinopsis infundibulum*). Most measure only 2-3 inches (5-7$^1/_2$ cm), but this comb jelly has been known to reach 6 inches (15 cm) in height. They are found in relatively shallow waters from the Arctic to the Gulf of Maine.

Also common in the North Atlantic is the Sea Gooseberry (*Pleurobranchia pileus*). This ctenophore (which grows to only about an inch in body length) trails two long tentacles covered with sticky filaments that serve to capture small planktonic creatures. When the Sea Gooseberry retracts its tentacles, the prey is wiped off into its mouth. This animal is found from Maine to Florida in shallow water near shore. Its swarms can total thousands of individuals.

*The most abundant animal on Earth, the copepod ▶ (photographed here through a microscope), is about half the size of a grain of rice. This tiny crustacean is one of the prime foods of all ocean animals which eat plankton. This is **Calanus finmarchicus**.*

Chain siphonophores ▶
are colonies of many
individual animals
floating together as a
single creature. They
are extremely fragile
and will break upon
the slightest contact.

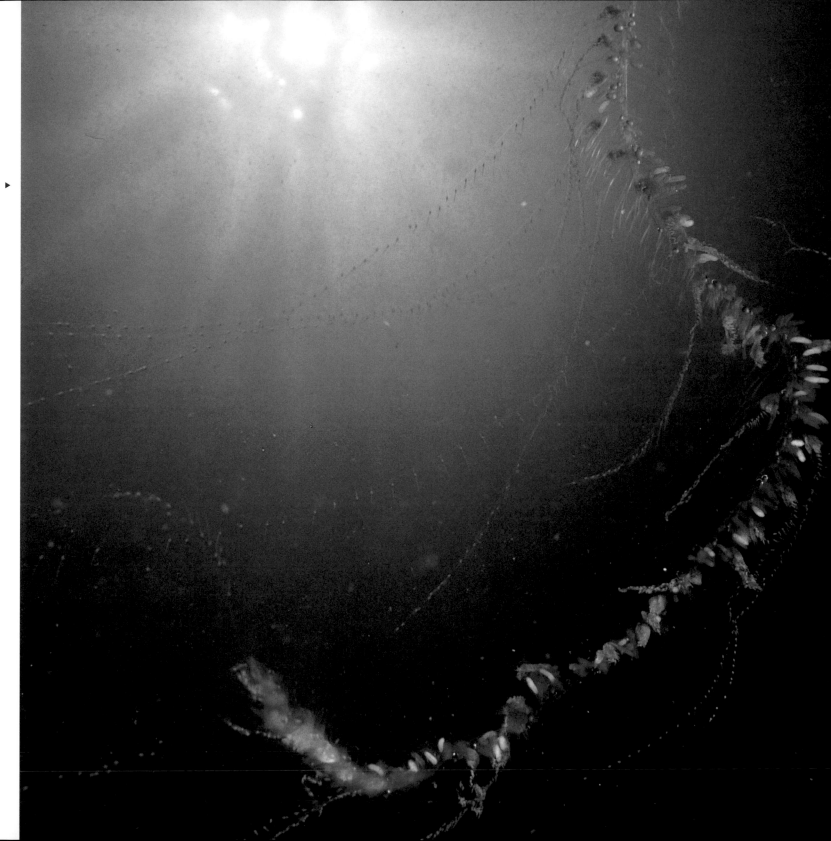

◄ *A Ctenophore, the*
Sea Gooseberry
*(**Pleurobranchia***
***pileus**). Ctenophores*
look like jellies, but,
in fact, have no stinging
nematocysts. Instead,
they capture food with
sticky filaments, seen
here trailing behind
the animal.

BENTHIC CREATURES

Benthic creatures live on or in the ocean floor, and the group is a large one. It takes its name from the Greek word meaning 'depth.' Of the 200,000 known marine animal species, about 98 percent, or 196,000, are benthic.

Benthic creatures are divided into two subgroups: those that live buried in the sand or mud (*infauna*) and those that live on top of or attached to the substrate (*epifauna*). Some benthic creatures are characterized as *sessile* and attach themselves to the substrate. For example, a hermit crab is benthic epifauna, but a sponge is sessile benthic epifauna.

The benthic environment is divided into two provinces, or regions. The region from the shore to a depth of about 650 feet (about 200 m, or the average maximum depth of the continental shelf) is called the neritic province. The suboceanic province includes depths beyond 650 feet (see figure 1).

Most benthic sea creatures live in the euphotic zone of the neritic province, where food is most readily available. Although some of the creatures I describe in this book can be found in the suboceanic province, a scuba diver is limited to the neritic province, where the photographs for this book were taken.

Most of the creatures seen here are commonly found within 100 feet (30 m) of the surface, and a few are even found above water at low tide. The tides are of great importance to many species, and an understanding of the tidal zone is important.

TIDES

Ocean tides result from the gravitational pull on the Earth of other celestial bodies. Although such Earth-bound forces as wind do affect the tides, the moon and the sun have by far the most powerful impact. When the moon and the sun are in alignment with each another, they exert tremendous gravitational pull on the oceans (see figure 2). We experience the results in tides that are extremely high or extremely low, a condition called a spring tide. These tidal extremes occur twice a month and are not related to the spring season.

When the moon and the sun are at right angles to each other with respect to the Earth (90° out of phase with each other, or at quadrature) their gravitational forces are minimized and result in weaker neap tides (see figure 3). Neap tides are not as high as spring high tides, nor are they as low as spring low tides. Therefore the tidal range, or difference in height between high and low tides, is much more extreme during a spring tide than during a neap tide.

North Atlantic tides are called semidiurnal; there are two high tides and two low tides every day. Because the moon orbits the Earth every $29\frac{1}{2}$ days, it is in line with the sun twice every $29\frac{1}{2}$ days and in quadrature twice every $29\frac{1}{2}$ days. Those alignments produce two spring tides and two neap tides every $29\frac{1}{2}$ days, roughly every month. Other factors also play a part in the tidal cycle, as, for example, the shape of a bay or location where the

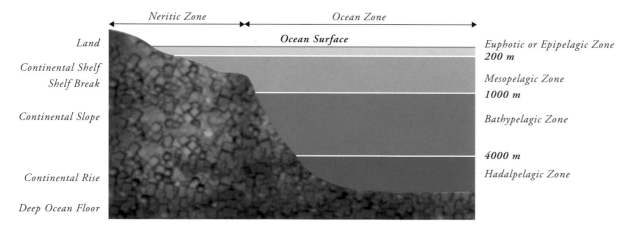

Figure 1 **Oceanic Provinces and Sea Floor Topography**

Spring Tide occurs twice per 29.5 days (when Sun, Moon and Earth are aligned)

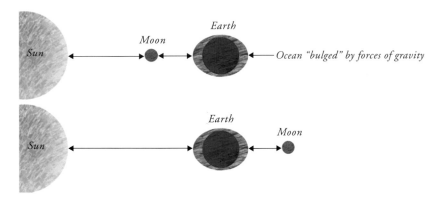

Figure 2 **Spring Tide**

Neap Tide occurs twice every 29.5 days (when Sun and Moon are in quadrature)

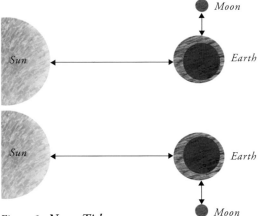

Figure 3 **Neap Tide**

tides are being measured. Many places, like the Gulf of Mexico, have diurnal tides (only one high and one low tide each day), while places like the Caribbean islands experience such small ranges that the tide is practically nonexistent.

Tidal ranges in the North Atlantic can be large, especially during a spring tide. In fact, the northern tip of Canada's Bay of Fundy has the greatest tidal range in the world: 56 feet (17 m) during a spring tide. That change is caused, in part, by a standing wave that results from the extreme length and narrowness of the bay. Most of New England experiences a 10-20 foot (3-6 m) tidal range that is enough to move large amounts of water in and out of New England bays. This frequent change of water makes many areas on the New England coast excellent sites for aquaculture (fish farming).

The tidal (*littoral*) zone is divided into four regions: the low-tide zone, the middle-tide zone, the high-tide zone, and the supralittoral, or spray, zone (see figure 4). The low-tide zone is the region between the lowest spring low tide and the highest neap low tide. The high-tide zone is the region between the highest spring high tide and the lowest

neap high tide. In between these two regions is the middle-tide zone. The supralittoral zone is a region above the highest high tide, which is never actually under water, but which receives a lot of seawater spray from the surf. Creatures may live in several of these zones but are not usually found in all four. For example, periwinkle snails, common on New England's rocky shores, are found mostly in the

spray and high-tide zones. Sea urchins and sea stars are found in and below the low-tide zone but not above it.

As we examine different creatures of the North Atlantic, you may be surprised at the remarkable number of animals that inhabit the littoral zone, living just beyond our sight as we stand watching the ocean waves rolling into shore.

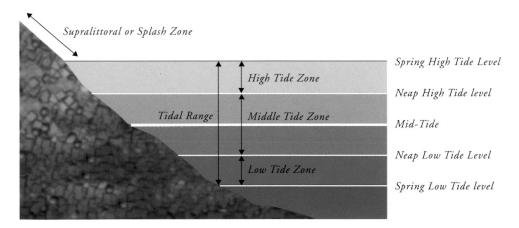

Figure 4 **The Tidal Zone**

PORIFERA

 SPONGES

Although they may look plantlike, sponges are among the simplest multicellular animals. A sponge is a sessile, benthic creature that attaches itself to something solid in a place where it may receive enough food to grow. The sponge phylum Porifera literally means "pore-bearing." A sponge is covered with tiny pores, called ostia, which lead internally to a system of canals that eventually connect to one or more larger external holes, called oscula. The canals of the sponge are lined with specialized cells called choanocytes, or collar cells. The collar cells have a sticky, funnel-shaped collar and a hairlike whip, called a flagellum. The collar cells serve two purposes. First, they beat their flagella back and forth to force water through the sponge. The water brings in nutrients and oxygen while it carries out waste and carbon dioxide. Second, the sticky collars of these cells pick up tiny bits of food brought in with the water. Amebocyte cells deliver food to other cells within the sponge. Sponges are effective filter feeders because they can capture and eat particles as small as bacteria, as well as much larger particles.

◄ *A small strangely shaped sponge,*
Melanchora elliptica.

The "skeleton" of the sponge is composed of either tiny needlelike splinters called spicules or a mesh of protein called spongin, and sometimes a combination of both. Many sponges can be identified only by a microscopic examination of the skeleton.

Most sponges are hermaphroditic (carrying both sexes) but produce only one type of gamete for each spawn, that is, some play the male role while the other plays the female role, even though they are both capable of playing either role. Sperm is released into the water column by "male" sponges and finds its way to "female" sponges, where fertilization occurs internally. Eventually, larvae are released from the female sponge and float around in the water column as plankton for a few days. Then they settle down, attach themselves to a spot, and start growing. Sponges may change sexual roles each time they reproduce.

There are many different types of sponges in North Atlantic waters, and they can be quite colorful and beautiful. Although not scientific classifications, it is useful to know that sponges come in two basic types: encrusting and freestanding. Two encrusting sponges common in New England are the Slime Sponge (*Halisarca dujardini*)

and the Purple Sponge (*Haliclona permollis*). These sponges typically cover the surface of a rock in much the same way that moss covers a rock on land.

Freestanding sponges are more interesting than encrusting sponges, but they are harder to find. Freestanding sponges have more interior volume compared with their outside surface area and therefore must live in a fairly nutrient-rich area in order to grow large.

The Finger Sponge (*Haliclona oculata*) is named for its shape and can grow to 18 inches (46 cm) high under ideal conditions. It is tan to brown-colored (although sometimes it is pale pink or lavender) and is found from Labrador to North Carolina.

The Crumb-of-Bread Sponge (*Halichondria panicea*) gets its name from its texture. Its color ranges from yellow to green, and it grows as large as 12 inches (30 cm) across and several inches high. Like many sponges, the Crumb-of-Bread has a pungent odor that probably protects it from predation. It is found from shallow water to depths greater than 200 feet (61 m) and from the Arctic to Cape Cod. This sponge grows to gigantic proportions in places of extreme tidal fluctuation, like Passamaquoddy Bay and the Bay of Fundy.

◄◄ *The Finger Sponge,*
   Haliclona oculata*.*

◄ *A unidentified,*
   free-standing
   North Atlantic
   Sponge.

17

CNIDARIANS

*◄ The Solitary hydroid (**Hybocodon pendula**) lives by itself on sandy bottoms, where it bends over and drags it tentacles through the sediment to catch food.*

Creatures of the phylum Cnidaria are among the simplest of the so-called higher organisms, but they are also among the most beautiful. All cnidarians, approximately nine thousand living species worldwide, are radially symmetrical. That means that parts of the body extend outward from the center like the spokes on a bicycle wheel. The sea star (a member of the phylum Echinodermata; see chapter 6) exemplifies the concept of radial symmetry.

Cnidarians include the hydroids, jellyfish, anemones, and corals. All cnidarians have tentacles with stinging cells in their tips that are used to capture and subdue prey. In fact, the phylum name *cnidarian* literally means 'stinging creature.' The stinging cells are called cnidocytes and contain a structure called a nematocyst. The nematocyst is a coiled threadlike stinger. When the nematocyst is called upon to fire, the thread is uncoiled and springs straight. The harpoonlike thread punctures

the cnidocyte wall to strike its prey. Most stingers release a toxin that helps disable the victim. The nematocyst is fired when the tentacle touches something or when prompted by nerve impulses from the animal.

Most cnidarians are not harmful to humans, since the stinger cannot penetrate human skin deeply enough to inflict any harm. Some jellyfish can, however, deliver extremely painful and, in a few cases, even fatal stings to humans.

The cnidarian can have one of two basic body types: polypoid and medusoid. Corals and anemones are polypoid. They have tentacles and mouths facing up and the other side is affixed to a substrate or connected to a colony of other creatures of the same species. Medusoid cnidarians live with their mouths and tentacles pointed down and are usually free-swimmers, like jellyfish. Thus the anemone is sometimes called an upside-down jellyfish, which is just about what it is!

 # CNIDARIANS

 ## HYDROZOANS

Although there are some 2,700 members of the class Hydrozoa (meaning 'water-animal'), most hydroids are small and are often mistaken for plants or go entirely unnoticed. One important and unusual feature of hydrozoans is the strange way in which many of them reproduce. Although the "adult" form of a hydrozoan is frequently benthic, many of these animals have a hydromedusa stage, during which they swim in the water column as tiny jellyfishes, or jellies. In New England, there are so many different hydroids that it would be impossible to discuss them all, but we can touch on some of the more common types.

The Solitary Hydroid (*Hybocodon pendula*), also known as the One-Armed Jellyfish, looks like a tiny sunflower. After a medusa stage, it matures and settles to the bottom to grow. During this polypoid stage, the Solitary Hydroid uses its long tentacles (usually about thirty in number) to search the sea bottom for food. To reproduce, it releases either sperm or eggs, depending on its sex. Within a cluster of these creatures on the sea floor, eggs and sperm will mix together in the water, and the eggs become fertilized. The fertilized eggs develop into larvae, called planulae (singular: planula), which become tiny jellies, and the cycle begins again.

Solitary Hydroids grow to about 4 inches (10 cm) high, and the circle of tentacles gets to be about $^3/_4$ of an inch (18 mm) across. The hydromedusa (juvenile) form is only about $^1/_4$ inch (6 mm) across. The species occurs from the Arctic to Long Island Sound.

The Tubularian Hydroid (*Tubularia crocea*) grows in colonies of up to several hundred individuals. Each polyp can reach a height of 5 inches ($12^1/_2$ cm) above the substrate on a long stem and has twenty-four large tentacles. Colonies can be 12 inches (30 cm) in diameter. Reproduction is different from that of the Solitary Hydroid. The Tubularian Hydroid has no medusoid stage but does produce a planula inside an "attached medusa" called a gonophore. The planula develops into an actinula (sort of a polyp with no colony) and is ejected into the water. The actinula settles down wherever the current deposits it, and a new colony begins to grow. The species occurs in shallow water from Nova Scotia to Cape Hatteras.

Some hydroids have even more complicated life cycles. Species in the genus Obelia live in large colonies. The colonies grow by means of simple asexual budding. To produce new colonies, the process is a bit more complicated. First, the colony

◄ *Tubularian hydroids (**Tubularia crocea**) grow in colonies of up to several hundred individuals.*

▲ *The Frilled sea anemone, **Metridium senile** . Although it looks like a flowering plant, this delicate creature is actually a plankton-eating animal.*

◄◄ *A Diver with group of Frilled sea anemones* (**Metridium senile**).

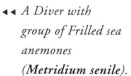

◄ *A group of orange colored* **Metridium senile**. *Coloration in this anemone ranges from completely white to brown, also including orange, pink and mottled patterns*

▲ *Variations in coloring of the beautiful Northern Red* ▶
*anemone (**Tealia crassicornus**). These are both the*
same species, although you might never suspect it
due to the large variations in color from individual
to individual.

◀ *Mouth of a Metridium anemone: the last*
thing a copepod ever sees!

buds off tiny jellies. The jellies then swim away and
release eggs and sperm into the water to form a
zygote. The zygote develops into a planula that
settles to the bottom and finally grows into a new
colony. Why all this effort just to start a new
colony? The process allows jellies and planulae to be
carried by the current before they settle and expands
the range of the hydroid population.

Not all hydrozoans are sessile benthic creatures. Just as the Solitary Hydroid has a medusoid
stage, some hydrozoans spend most of their life-
times as free-swimming medusoids, or jellies. One
well-known example is the Portuguese Man-of-War
(*Physalia physalis*). This jelly, although not a true
jelly (that is, in the jellyfish class Scyphozoa), is
known scientifically as neuston. While plankton
drifts and nekton swims, a neuston "floats." The
Physalia floats on the surface and uses a saillike
appendage on its body to catch the wind.

The *Velella velella*, commonly called a By-
the-Wind Sailer, is similar to the *Physalia*. These
neuston are found in all the worlds oceans. The
Velella is harmless to humans, but the *Physalia* is
capable of delivering powerful stings and burns,
even after it is dead. It contains one of the most
powerful venoms known in marine animals.
Furthermore, its tentacles can be as long as 60 feet
(18 m), so divers can be stung by this creature
without ever seeing the body of the animal. While
the Velella is a single hydrozoan, the *Physalia* is
actually a colony of many hydroids living together
as a community and acting as an individual.

More variations in coloring of the ▶
beautiful Northern Red anemone ▶ ▶
(Tealia crassicornus).

 ANTHOZOANS

The term *Anthozoan* literally means 'flower-animal.' The only thing anthozoans have in common with flowers, however, is their appearance. This class includes the beautiful anemones and corals. In the North Atlantic, there are several different types of anemones and even a few corals.

 ANEMONES

An anemone is physically similar to a hydroid, but it does not grow in colonies. It is a solitary polypoid creature that attaches itself to a substrate of some sort and waves its tentacles in the water to catch prey. The tentacles contain stinging nematocysts and surround a mouth leading to a gastric cavity. The anemone has no anus and rids itself of waste by regurgitating. Unlike hydroids, however, anemones have a muscular body, called a column. The animal can use both lateral and longitudinal muscles to contract or extend itself. Thus, when threatened the anemone can shrink and hide by closing up and forcing water out of its body.

The Frilled Anemone (*Metridium senile*) is relatively large and can be found in several color schemes, including white, brown, orange, and white with brown spots. When completely retracted, the animal looks rather like a bagel. When extended, however, it displays a large quantity of thin ten-

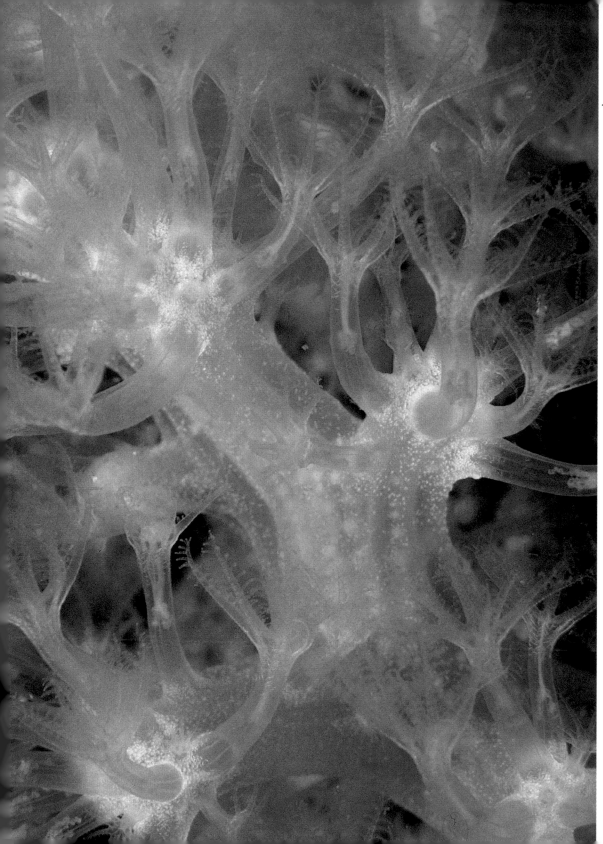

◄◄ *A non-branching type of coral sometimes called Dead Man's Finger coral (**Alcyonium digitatum**).*

tacles. The Frilled Anemone, unlike some cnidarians, develops more tentacles as it gets larger. Large examples can have as many as one thousand tentacles, while small ones may have fewer than one hundred. The anemone attaches itself to rocks with its pedal disk. Although the animal can and does reproduce sexually, it can also reproduce asexually by a process known as pedal laceration. The anemone is not stationary but can creep along the rocks to which it attaches. Frequently, it scrapes off small pieces of its pedal disk (the part of the anemone that is in contact with the substrate) as it moves. These tiny chunks of tissue, which may be less than 1/16th of an inch (1.5 mm) wide, will produce tentacles and begin to grow within a week to two weeks. It is not uncommon to see a trail of small anemones sprouting up along the path of a larger anemone's migration. Frilled Anemones are the most prolific in New England and are common from the Arctic to Delaware. They are also found in the Pacific Northwest. The animal can reach 18 inches (46 cm) high and 9 inches (22 cm) in diameter. It consumes very small zooplankton, such as tiny amphipods, copepods, and fish fry. In the North Atlantic, Frilled Anemones feed primarily on the larvae of barnacles and mussels. Their nemato-cysts are harmless to humans and larger fishes.

The Northern Red Anemone (*Tealia crassicornus*) is one of the most beautiful in the

◄ *Not all Red Soft Coral colonies are actually red. This close-up of a white colony shows the polyps, each containing eight tentacles.*

world, and I have searched widely for the many color variations of this cold-water anthozoan. Few things about the Northern Red Anemone are as they seem. They are not always red, for example. They are frequently white, orange, or pink but are also found in yellow, purple, and banded combinations of those colors. Because this unusual creature is frequently not red, many people prefer to call it the Tealia anemone – which is fine on the East Coast, but on the West Coast the term can be confusing because there are other species of anemone that share the genus (like *T. lofotensis*, the Strawberry Anemone, and *T. coriacea*, the Leathery Anemone).

The Northern Red Anemone has one hundred thick tentacles arranged in circular rows around its mouth. Large nematocysts allow the Tealia to capture even small fish. I watched once as a green sea urchin fell from an overhanging rock onto a large Northern Red Anemone. In a moment, the anemone enveloped the urchin in a deadly ring of tentacles from which there was no escape. A diver who places a finger among the tentacles of a Northern Red will be surprised at the degree of force the animal can exert to pull him in. Fortunately, this anemone, like most others, cannot sting through the skin of a human, or a diver's neoprene glove.

The Northern Red Anemone is found on the east coast from the Arctic to Cape Cod Massachusetts, and on the west coast from Alaska to northern

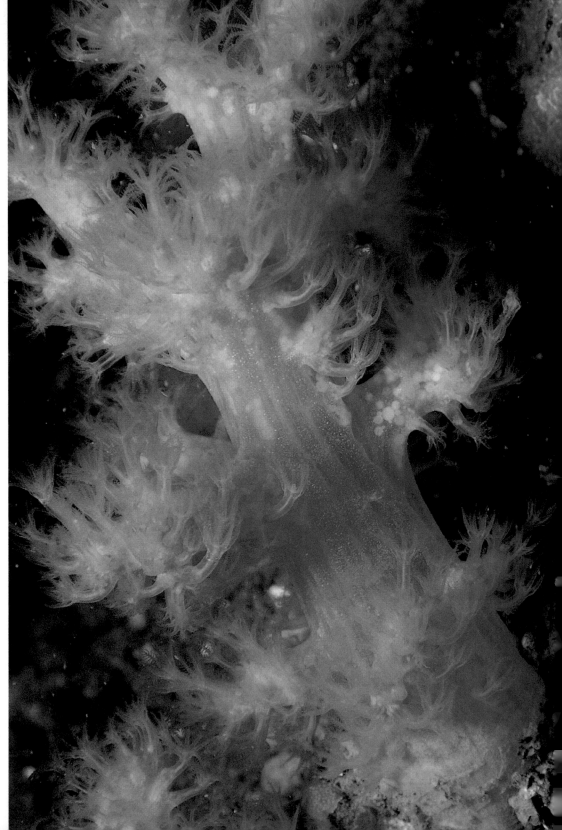

Looking much like its Indo-Pacific cousins, this stunning Red soft coral ▶
*(***Gersemia rubiformis***) is found in the frigid waters of the North Atlantic
from the Arctic to the Gulf of Maine.*

 # CNIDARIANS

California. In Massachusetts, examples are relatively rare and found only in deep water, where the temperature stays below 50°F or so all year around. Northern Maine, on the other hand, has so many that in places they carpet the bottom of the sea. These anemones thrive (as do most others) in areas where strong currents or large tidal flows refresh the supply of food-laden water.

Some reference books report that the Northern Red Anemone grows to only 3-4 inches (7-10 cm) in diameter. Although I have never brought a ruler along on a dive, I have photographed examples as large as 8 inches (20 cm) across.

The first time I saw a Northern Red Anemone, I was led to it by a fellow diver I had met at a popular dive spot on the Rockport, Massachusetts, shore. He was rinsing his camera after the first dive of the day, just as I was getting out of the water from my first dive. The sun was shining and the ocean was calm. We did not know one another, but underwater photographers are so uncommon in New England that it is natural to stop to chat. He told me that he was photographing red anemones. I had never seen one off Rockport, but my new friend assured me that they were there and offered to lead me to them on his next dive.

After lunch, we began our dive by snorkeling at least a hundred yards (90 m) from shore. Since the bottom drops off fairly quickly, I was beginning

to wonder whether I was diving with a madman. "Are you sure that they're out this far?" I shouted, while struggling with my cumbersome camera system. "Trust me," he replied, "We're over them right now."

We prepared to submerge, let the air out of our buoyancy compensators, and started sinking. I followed him down to 30 feet (9 m), where the water was getting both darker and colder, but the bottom was still out of sight. We passed 60 feet (18 m), where the pressure is almost three times greater than it is at the surface, and the temperature was 45°F, 20° colder than the water above. The bottom finally came into sight at 70 feet (21 m). We touched down at 80 feet (24 m). It was very dark and colorless.

Water is a light filter. Blue and green penetrate water fairly well, but longer wavelengths of light – red, orange, and yellow – get filtered out at greater depths. In 80 feet of water a red anemone looks gray, and there is no way to see its color without a light. The experience is like watching television in black and white. In the North Atlantic everything appears in shades of the same grayish-greenish color.

Using his flashlight, my guide pointed out a beautiful Northern Red Anemone. It was deep red, but looked gray and colorless without the light. I had never seen one before, for two reasons. First, I

had never dived deep enough, where the water is really cold all year around, and, second, without using a light to bring out color, it is possible to swim past a large Tealia and not see it. I would later dive among thousands of these anemones in rather shallow water (where the colors stand out better) in northern Maine, but a light is still important if one is to see the true color of an anemone, or any other creature, underwater.

Not all anemones attach themselves to hard surfaces like the anemones in the Actinarian order do. Anthozoans in the order Ceriantharia are adapted for life on muddy or sandy bottoms. Sometimes called tube-dwellers, or burrowing anemones, they secrete a fleshy tube anchored in the sand or mud into which the animal can retract itself. This tube is made of mucous and threads fired from ptychocytes, organelles similar to nematocysts.

In the North Atlantic, the Northern Cerianthid (*Cerianthus borealis*) is a common burrowing anemone that looks small at first glance. Its two rings of tentacles reach a diameter of only $1\frac{1}{2}$ inches (4 cm), but the Northern Cerianthid produces a long tube extending to depths of 18 inches (46 cm). This anemone is found from the Arctic to Cape Cod and from the shallows to depths greater than 1,300 feet (400 m).

CNIDARIANS

CORALS

Tell people there are corals in the cold North Atlantic Ocean and they are skeptical. Corals are anthozoans that form colonies. On close examination, each polyp in the colony resembles a tiny anemone. So-called stony corals secrete a limestone (calcium carbonate) base, which grows over time into a coral reef. Other types of coral do not build reefs and exist in smaller colonies. One particularly beautiful coral in the Order Alcyonacea within the subclass Octocorallia is known commonly as soft coral. Its scientific name comes from the eight tentacles on each of its polyps. Octocorals resemble plants since they are flexible and resemble small trees. Soft corals secrete calcium carbonate in the form of limestone spicules (needles) that are embedded in the colony to give it strength while allowing the colony to bend in a current.

The North Atlantic has a stunning soft coral known simply as Red Soft Coral (*Gersemia rubiformis*). It is found from the Arctic to as far south as the Gulf of Maine. Although this coral looks exactly like the tropical soft corals of the South Pacific and Red Sea, it is much smaller, reaching a height of only 6 inches (15 cm). Like the Northern Red Anemone, this coral is not always red. It is found in pink, orange, and white as well as red.

After learning about these corals, I ventured on a search for them with Oceanic Research Group diver and cinematographer Tom Krasuski. We traveled to northern Maine, where the weather is unpredictable and the water very cold. On our first dive of the search, we sank down slowly through dark, numbingly cold water, our eyes straining for a glimpse of anything vaguely resembling coral. At most, I had hoped to find just tiny clumps, if any at all. To my surprise, as we passed the 40-foot (12-m) level the bottom was absolutely covered with breathtaking Octocoral. We returned to the site of this first dive several times before heading home. My logbook indicates that I have since been to this spot more than fifty times. Each dive reveals something new.

The soft coral photographs pictured here are from that first dive site. We returned to northern Maine three times that summer for further exploration. We found those cold northern waters teeming with marine life. Much of this book was shot there, and it has become one of my favorite places on earth to dive.

COMB JELLYFISH

Comb jellyfish were formerly grouped in with cnidarians in a phylum called Coelenterata. Since comb jellies do not have nematocysts, but all other coelenterates do, comb jellies have now been assigned their own phylum, Ctenophora. There are some fifty species of ctenophores (see Chapter 1).

◄ *The Lined anemone (**Fagesia lineata**), growing a little over an inch high, is frequently found in patches of great numbers, though it is not colonial.*

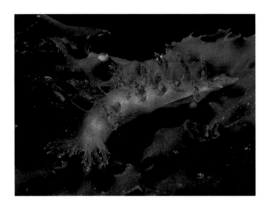

MOLLUSKS

Mollusks are among the most familiar invertebrate sea creatures. Though snails, clams, mussels, squid, and octopods might not seem to share obvious physical characteristics, they are, in fact, remarkably similar. There are more than fifty thousand species of mollusks, which ranks the phylum second only to the arthropods in number.

Mollusks have three body regions: a head, a visceral mass, and a "foot." The head contains the sense organs and "brain," while the visceral mass contains the internal organs. The "foot" is the muscular lower part of the body, which is in contact with the substrate. Mollusks usually have a shell – but not always – and also an extension of the body wall called the mantle, which secretes its shell. The mantle encloses a cavity that contains the gills, anus, and excretory pores.

Many mollusks have a radula, a sort of rough, sandpaper-like tongue, which is used to scrape up food. The radula is made of a hard material called chitin, the same material of which arthropod exoskeletons are made. In addition, some mollusk radulae are impregnated with magnetite to give them superior wear characteristics.

Mollusks have well-developed body organs (nervous system, circulatory system, respiratory system, and so forth), but their body is not segmented. There are seven classes of mollusks. I discuss Gastropoda, Polyplacophora, and Bivalvia here and will describe the Cephalopods (squid and octopods) later. Aplacophora, Monoplacophora, and Scaphapoda are rare or extremely deep-water creatures, and I do not cover them here.

THE GASTROPODS

The class Gastropoda (meaning 'stomach-foot') contains about 70 percent of the molluscan species and includes such familiar animals as snails, limpets, nudibranchs, and abalones. Snails, limpets, and abalones have a shell, while slugs and nudibranchs do not. There are a small number of land gastropods (that is, some snails, slugs, and so on) that I do not discuss here.

 # MOLLUSKS

 ## THE SNAILS

Of the many varieties of North Atlantic snails, there are several species that are fairly representative of the group. One of the most common snails on the North Atlantic's rocky shores is the periwinkle. Low tide reveals these animals, which forage for algae growing on the rocks near the high-tide line. As the tide ebbs, periwinkles seek shade, retract into their shells, close the little trap door to the opening of their shell (called an operculum), and wait for the water to rise again.

The common periwinkle (*Littorina littorea*) arrived on the North American continent only little more than a hundred years ago, probably by stowing away on European ships. They are now so common in the intertidal zone that it is hard to imagine the North Atlantic sea coast without them. When the pilgrims arrived in the New World, however, they found no periwinkles on this side of the Atlantic.

Moon Snails (*Lunatia heros*) are notable for their enormous size. These giants frequently reach 5 inches (12^1/$_2$ cm) in length and live fully submerged on sandy ocean bottoms. The Moon Snail produces a large mantle that can completely envelop the shell by filling it with water. The shell grows in segments, and a new section is easily distinguished from an older portion by its different color and weathering. The entire mantle and foot can be withdrawn into the shell for protection.

◄ *The egg case of the moon snail resembles an old piece of broken pottery. Here a Rock crab (***Cancer irroratus***) has taken refuge from the camera in the first thing it could find on the sandy bottom.*

Moon Snails like to burrow. I once put two small Moon Snails into a marine aquarium to watch their behavior. They dug into the sand almost immediately and were out of sight for weeks. I was convinced that they were both dead, until I observed the tank late at night. Both were cruising the bottom of the tank eating excess fish food that had settled there. Although one fell prey to a hungry Northern Sea Star, the other continued making a nocturnal food run each night for years.

Night dives in New England over sandy bottoms usually reveal Moon Snails feeding, although they may be seen during the day. It is interesting that certain bays and coves are home to many Moon Snails, while adjacent locales rarely have any. This difference may be related to the availability of food and the proximity of predators.

Moon Snails lay eggs in clusters called sand collars. At a distance, sand collars resemble old pieces of broken pottery. Closer inspection reveals the sand collar to be flexible. Thousands of eggs in the collar are glued together with a sticky mucous that fish find distasteful, which is a protection against predators. Many first-time scuba divers have reached out for a sand collar thinking it was discovered treasure, only to find the floppy sand collar. At night, hermit crabs and rock crabs frequently hide in sand collars as divers approach. Although the collar offers no real protection, the behavior is humorous to watch.

Whelks are another large-shelled snail common in the North Atlantic. They are similar in appearance to the conchs found in tropical seas but are smaller. They have a coiled shell into which they can retract, as well as an operculum to seal the mouth of the shell. Whelks usually have a probiscus used to explore nooks and crannies for food. A whelk will crawl along the ocean floor searching for food, which often turns out to be a dead creature. Most whelks are carnivorous.

Often whelk shells are ornate and have many points and ribs (probably for defense) in contrast to the smooth shell of the periwinkle.

 ## LIMPETS

Limpets share the snail's anatomy but lack a spiral shell into which it can retract. Limpet shells resemble a flattened cone or a volcano. For protection, it must clamp down hard against a surface. Breaking a limpet free from a rock once it has clamped down is nearly impossible.

The only limpet common in the North Atlantic is the Tortoise-Shell (*Notoacmaea testudinalis*), found from the Arctic to Long Island Sound. This gastropod reaches about 1^1/$_2$ inches (<4 cm) in length and prefers clinging to rocks in shallow water or even above the low-tide line. It can clamp down so tightly to the substrate that attempting to pry it loose will break its shell.

 # MOLLUSKS

 ## THE NUDIBRANCHS

Nudibranchs are coveted by underwater photographers because they are both beautiful and rare (underwater photographers always enjoy a good challenge). Every ocean and every region within those oceans contains nudibranchs, although they can be difficult to find. These creatures are essentially snails without shells. They resemble garden slugs, from which they take their common name: sea slugs. Unlike garden slugs, however, nudibranchs can be brilliantly colored. This coloring may be *aposematic* – intended to warn potential predators that they taste bad.

The North Atlantic is home to more than thirty nudibranch species. Many are rare or extremely small, while others are so similar to one another it is difficult to distinguish them. The term *nudibranch* literally means 'naked-gill.' Unlike snails, whose shell covers and protects their gills, the nudibranch's gills (when present at all) are exposed.

There are two common body styles of nudibranch: the Dorid and the Eolid. Both types absorb oxygen from sea water through their skin, but the Dorid nudibranch has a tuft of gills (ctenidia) on its rear dorsal surface to assist in respiration. It can usually retract these gills to protect them.

◄ *The Waved whelk (***Buccinum undatum***) is a scavenger of dead animal matter, using its long probiscus to reach into carcasses. The white spotted material is the mantle, while the brown disk attached to the mantle is the operculum, used to cover the opening of the shell when the animal retracts.*

The Eolid does not have a specific cluster of gills. Instead, it has many pointy extensions on its dorsal surface. These extensions, called cerata, resemble the fine tentacles of an anemone and serve two purposes. First, they significantly increase the surface area – and oxygen absorption – of the nudibranch's skin. They are sometimes erroneously referred to as gills. Second, the cerata serve as defensive weapons. Many Eolid nudibranchs dine on a variety of hydroids and anemones. The nudibranch eats the cnidocytes of its prey and swallows them without discharging the venomous nematocysts. It places the nematocysts in the tips of the cerata, where they function as stinging anemone. Without cnidarian prey to supply nematocysts, the nudibranch would be defenseless.

The Dendronotus is a third kind of nudibranch found in the North Atlantic. Like an Eolid, the Dendronotus has cerata, but they are branched like the arms of a basket star or sea cucumber and do not hold captured nematocycts. While serving as camouflage, they also provide increased surface area for respiration.

All nudibranchs are hermaphroditic. This does *not* mean that they reproduce asexually; reproduction still requires two individuals. Hermaphroditic reproduction is nature's way of increasing the likelihood that two nudibranchs will meet, be compatible, and produce offspring. For the nudibranch, whose mobility is limited to a slow crawl and whose population density is low, hermaphroditic reproduction is important to survival.

▼ *Limpets are basically just flattened snails with volcano-shaped shells. The Tortoise-Shell Limpet (***Notoacmaea testudinalis***) can clamp down against a rock with such tenacity that it cannot be pried loose without breaking its shell.*

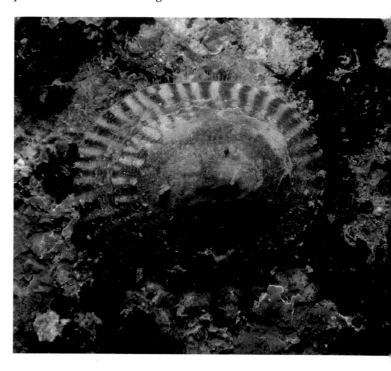

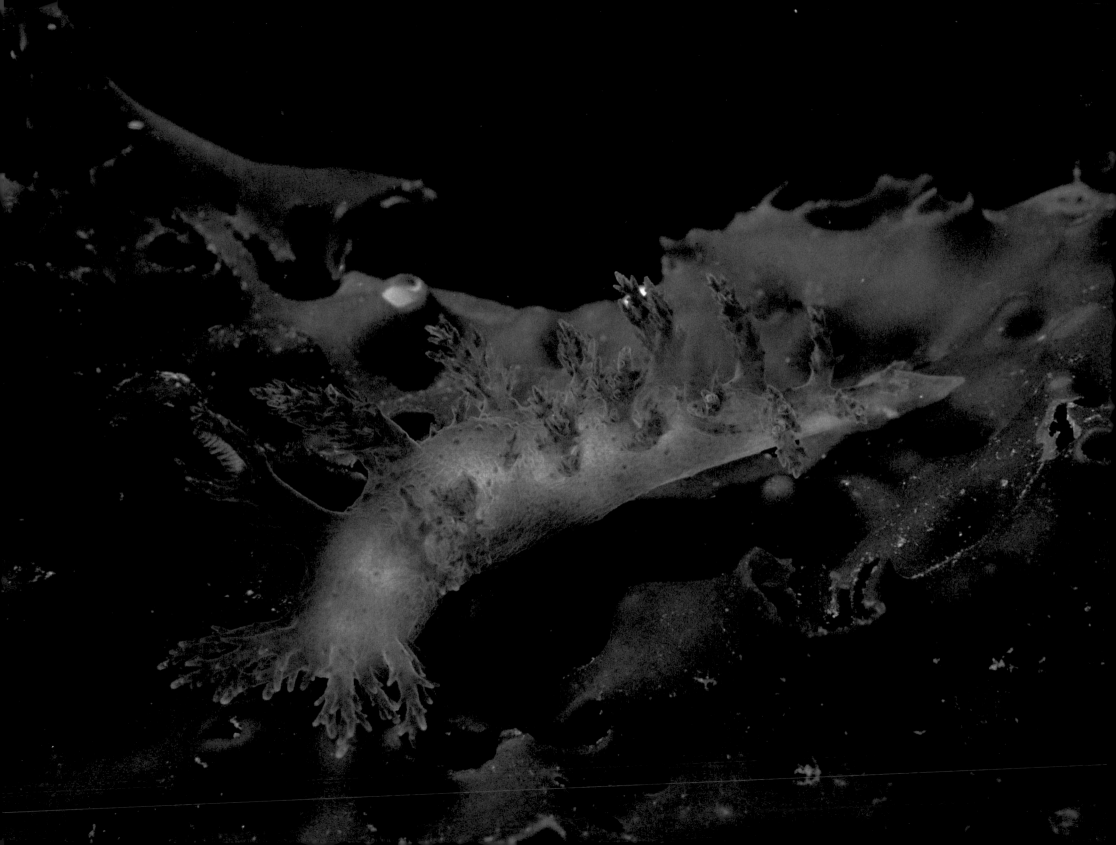

 # MOLLUSKS

 ## THE DORIDS OF THE NORTH ATLANTIC

Among the Dorid nudibranchs in the North Atlantic, one of the most beautiful is the Cadlina (*Cadlina laevis*). It is oval-shaped and white, with brilliant yellow speckles and a yellow perimeter. Extremely shy, it can detect movement in the water so far away that photographing one with its gills exposed is difficult. This nudibranch eats encrusting sponges. As is the case with many nudibranchs, this one takes on the appearance of its favorite food and is sometimes difficult to spot while it is crawling along, especially since it is only about 1 inch (2.5 cm) long.

The Cadlina lays eggs in a spiral pattern on rocks during the early spring. Unlike most nudibranch eggs, which hatch into tiny platonic larvae and swim away, the Cadlina's eggs develop directly into tiny baby nudibranchs with no intermediate planktonic stage.

To protect itself, the Cadlina frequently feeds between cracks in rocks, making it less vulnerable to attack and difficult to photograph. It is discouraging to search – dive after dive – for a particular nudibranch, only to discover it hiding cleverly just out of photographic reach. (Dives like that lead one to wonder whether the nudibranch is actually smarter than we think. At the dive shop, they still tease me about my continuing search for the "elusive Atlantic nudibranch.")

Less pleasing to the eye, but no less interesting, is the Rough-Mantled Doris (*Onchidoris bilamellata*), named for its bumpy-textured outer surface. Brownish in color and about 1 inch in length, it prefers shallow water (usually fewer than 30 feet [9 m]) and is found from the Bay of Fundy to Rhode Island. The Rough-Mantled Doris dines on acorn barnacles and can be found in huge numbers in the early spring. As protection, it secretes sulfuric acid from glands in its dorsal surface, which makes the Rough-Mantled Doris an unpopular meal. The acid produced by Gulf of Maine residents is, on average, less concentrated than that produced by its Pacific and European cousins, perhaps because predation on the Doris in the Gulf of Maine is less intense.

◄ *The Bushy-Backed nudibranch (**Dendronotus frondosus**) has branched cerata on its back to help it blend in to its surroundings.*

 ## THE EOLIDS OF THE NORTH ATLANTIC

The Red-Gilled Nudibranch is among the most common found in the North Atlantic. Several species of this nudibranch share the common name Red-Gilled, including *Coryphella rufibranchialis* and *C. verrucosa*. Ironically, all are misnamed since Eolids have no gills. Actually, the cerata, numbering about one hundred, are red (or orange or brown). The Red-Gilled Nudibranchs grow to about $1^{1}/_{4}$ inches (3 cm) long and can be difficult to find, even though they range from the Arctic to New York.

The Red-Gilled Nudibranch prefers extremely cold water. In the northern latitudes, they are common all year around, but in the southern part of their range, they are found only in the fall through early spring. With the approach of warmer weather and water, the animal moves into deeper and colder water. Searching for this nudibranch in southern New England is fruitful only between February and June. Still small in February, the nudibranchs peak in population and size by May and become scarce by June. In August, seeing one is rare.

Eggs are laid in the late winter through early spring (February to April) on rocks and seaweeds. The Red-Gilled Nudibranch's egg mass resembles a tiny white ribbon. When the larvae are born, they are planktonic (swimming freely) and, strangely enough, have a shell. As the creatures grow, they shed their shells and adapt to a benthic existence.

The Salmon-Gilled Nudibranch (*Coryphella salmonacea*) is similar to the Red-Gilled Nudibranch

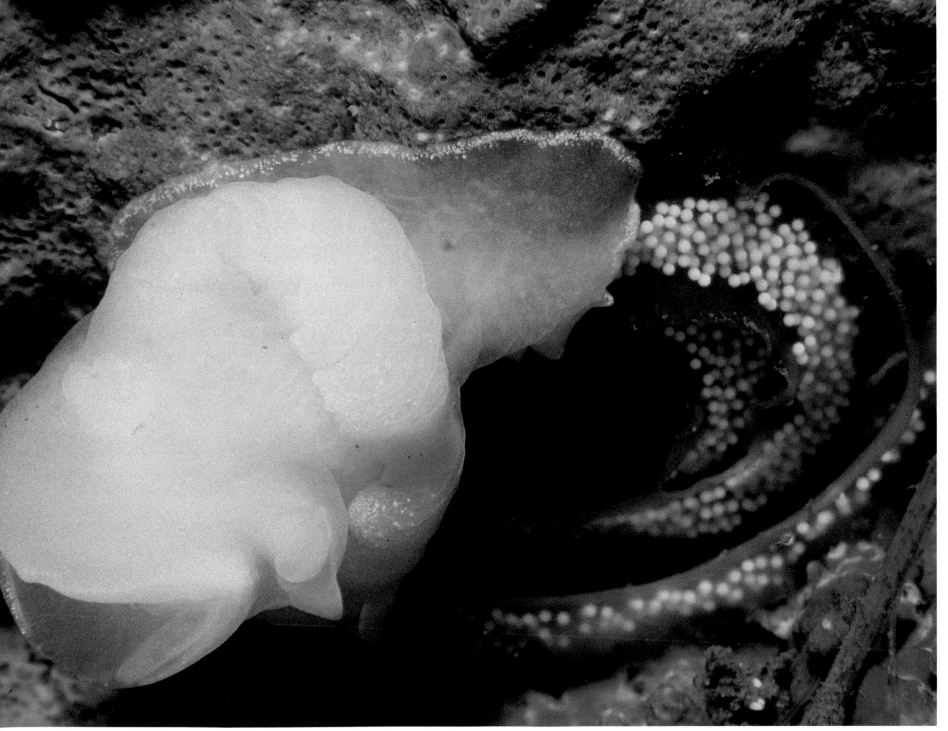

◄ The Cadlina
nudibranch laying
its eggs in a spiral.
Unlike most other
nudibranchs whose
eggs hatch into
planktonic larvae,
the Cadlina eggs
develop directly
into miniature
nudibranchs.

◄ *This beautiful Dorid nudibranch, called a Cadlina (**Cadlina laevis**), resembles
its favorite food, a sponge. The gill cluster is visible at the right.*

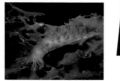 # MOLLUSKS

in body structure but grows slightly larger ($1^3/_4$ inches, or $4^1/_2$ cm, in length). It is a white-to-pink color. Like the Red-Gilled Nudibranch, this nudibranch has about one hundred cerata and feeds on hydroids. It prefers cold water and ranges from Greenland to Massachusetts. The Salmon-Gilled Nudibranch is similar to the Red-Gilled in its population distribution throughout the year.

The Maned Nudibranch (*Aeolidia papillosa*) grows to a substantial 4 inches (10 cm) long (fairly large for a North Atlantic nudibranch) and feeds on sea anemones, particularly the Frilled Anemone. It usually takes on the color of the anemone it is eating and, like the other Eolids, recycles the nematocysts of its prey. In exceptional cases, I have seen groups of these nudibranchs consume an entire anemone, leaving behind only a patch of clean rock. More commonly, the nudibranch attacks an anemone and is either driven off or gives up after it has had its fill. The anemone usually suffers no more than a superficial wound, which heals quickly. The Maned Nudibranch ranges from the Arctic to Maryland from the low-tide line to water more than 2,000 feet (610 m) deep.

In the North Atlantic, the most common Dendronotus nudibranch is the Bushy-Backed Nudibranch (*Dendronotus frondosus*). The species name *frondosus* (Latin for 'leafy') aptly describes this creature. The Bushy-Back's unusual "leafy" body

shape blends in well with the seaweed, where it hides and feeds. It can reach $4^1/_2$ inches (11 cm) in length and is found from the Arctic to New Jersey in depths between the low-tide line and water 350 feet (107 m) deep. They are usually brown, greenish, or purple in color and sometimes even white. We found a large colony of completely white nudibranchs feeding on hydroids in about 50 feet (15 m) of water in northern Maine. As is typical, a nudibranch takes on a color that camouflages it among whatever it prefers to eat. Purple might seem an odd color since there are few purple things in the ocean, especially things that a nudibranch will eat. In the dark green North Atlantic, however, purple looks brown, except in extremely shallow water. Because these nudibranchs are found in great abundance in Maine among brown and green seaweeds, purple is actually good camouflage.

CHITONS : THE CLASS POLYPLACOPHORA

Chitons are creatures that carry eight plates and outwardly resemble, but are not related to, the "pill bugs," which are commonly found under logs and rocks in the forest. Members of the class Polyplacophora (meaning 'many-plated'), Chitons crawl along rocks and forage for food (mostly algae)

using their radulae to scrape it off the substrate. The mineral magnetite hardens the radula of Chitons, so they can scrape coraline algae off rocks. These radulae are so hard that they can etch glass.

Extremely slow moving, a Chiton may not move more than 10 feet (3 m) in a year. They can detect the presence of light with their primitive eyes, which are embedded in the shell plates.

In the North Atlantic, three Chitons are common. The Mottled-Red Chiton (*Tonicella marmorea*) is found from Greenland to Massachusetts Bay. It grows to $1^1/_2$ inches (4 cm) long and likes to eat algae, sponges, hydroids, and bryozoans. Like all chitons, it is nearly impossible to remove it from a rock once it has clamped down, and it will curl up into a ball to protect itself if it is somehow removed.

The Northern Red Chiton (*Tonicella rubra*) is nearly identical in appearance to the Mottled-Red Chiton, except that it grows only about half as large ($^3/_4$ inch, or 2 cm, long). Both share nearly the same geographic range.

The White Chiton (*Ischnochiton albus*) is found from the Arctic to Massachusetts in water up to only 25 feet (7.5 m) deep. It thrives only in very cold water, making it common in the Bay of Fundy but rather rare in the southern parts of its range. It reaches only about $^1/_2$ inch ($1^1/_4$ cm) in length.

◄ *The White doris (**Onchidoris muricata**) is usually small ($^1/_2$ inch long).*

 # MOLLUSKS

THE CLASS BIVALVIA

The bivalves (meaning 'two shells') are perhaps the best-known mollusks because of their history as a source of food. Clams, mussels, oysters, and scallops are all bivalves. There are some fifteen thousand known species of bivalves, 80 percent are marine, while the balance are found in freshwater.

Bivalves may burrow through the sea bottom or attach themselves to the substrate with gluelike strings called byssal threads. Some scallops do not attach themselves; instead, when threatened, they prefer to swim away propelled by a squirt of water forced from their mantle.

Bivalves feed by filtering organic particles from the water and do not have a radula. A mucous coating on the gills traps food particles as water passes through them.

The shell of the bivalve is generated by the mantle from the inside. Pearls are made by clams, oysters, and mussels when a grain of sand or other small irritant becomes painfully stuck in the mantle of the creature. The bivalve coats the irritant with

◀ *A Red-Gilled nudibranch (Coryphella verrucosa) eats hydroids and stores the hydroid's stinging nematocysts in its cerata undischarged, so that it may use them to sting other creatures.*

the same material that is secreted to produce the inner lining of the shell. This makes the irritant smooth and, theoretically, less painful to the bivalve. Although many people think pearls come only from oysters, most bivalves can produce pearls, as can some snails, like the conch. The North Atlantic has innumerable species of bivalves.

THE SWIMMING SCALLOP

While most bivalves orient themselves vertically in the substrate, scallops generally orient themselves horizontally, with one shell in contact with the sea bottom and the other facing up. It is easy to tell which side was facing up when examining a specimen because the upper shell carries the remnants of all types of marine growth.

The scallop maintains this horizontal position with no attachment to the bottom so it can make a quick getaway in case of trouble. Usually, just closing its shell will protect the scallop from a predator. Sometimes, however, the situation calls for a tactical withdrawal, and the scallop can do that by quickly closing its shell and forcing the water from its mantle out of a nozzle. The scallop can thus make a jet-propelled escape. It is a remarkably successful maneuver. In the blink of an eye, the scallop is gone.

THE BRACHIOPODS

The creatures in the phylum Brachiopoda outwardly resemble bivalve mollusks and are frequently mistaken for them, but they are not related. Brachiopods (commonly called Lampshells) are bivalves and have two shells, just like clams and mussels, but their internal structure is unlike that of mollusks. The Brachiopod has tentacles, covered with cilia, between its shells. The beating of the cilia drives water over the tentacles, which gather small organic particles. Brachiopods also use a stalk to anchor themselves to the substrate.

Brachiopods have an interesting fossil history. Although only 280 species of Brachiopods now exist, there are more than 30,000 known fossil species, dating from as long ago as 600 million years. The record for notable age, though, goes to the Brachiopod *Lingula*, which is the oldest living genus of animal life on Earth. The *Lingula* dates back at least 425 million years.

The only common species of Brachiopod in the North Atlantic is the Northern Lampshell (*Terebratulina septentrionalis*). Growing to just more than an inch ($2^1/_2$ cm) long, it is found from the low-tide line to depths of more than 12,000 feet (3,658 m). It is widely distributed, from Labrador to New Jersey. Frequently found with sponges growing on its shells, the Northern Lampshell may use this symbiotic defense against sea stars, which cannot grasp the brachiopod when a sponge is in the way.

◄◄ *The Maned nudibranch (**Aeolidia papillosa**) chiefly consumes Frilled anemones. A group of these nudibranchs can actually consume an entire anemone! This large specimen is about 3 inches long.*

◄ *A pair of Maned nudibranchs (**Aeolidia papillosa**) mating. All nudibranchs are hermaphroditic, meaning that each individual contains both sexes. Reproduction still requires two animals, but any two individuals are compatible, and can mate.*

◄◄ *The Mottled Red Chiton (**Tonicella marmorea**) clamps down to rocks and gnaws away at algae. The teeth on its radula are hardened by magnetite, and are so hard that they can etch glass!*

◄ *A group of Blue Mussels (**Mytilus edulis**) with attached barnacles. These mussels sometimes occur in large numbers, and are edible, though not particularly commercially exploited.*

▲ *Although they resemble bivalve Mollusks, Brachiopods are not actually Mollusks at all, but are in their own phylum, and are frequently called "lampshells." These Northern Lampshells (**Terebratulina septentrionalis**) are covered with encrusting sponges, but their feeding tentacles are visible through their gaping valves.*

ARTHROPODS

◄ *The two dissimilar claws of the lobster have different purposes. The larger claw is a crusher, while the smaller claw is a cutter.*

The phylum Arthropoda includes more species and more creatures than any other group on Earth. There are nearly 1 million species of arthropods, and more than 90 percent of them are insects. Of the remaining 85,000 species, three include marine animals: the mostly marine subphylum Crustacea (30,000 species), the entirely marine class Pycnogonida, also called sea spiders (500 species), and the entirely marine class Merostomata, commonly called horseshoe crabs (5 species).

All arthropods share certain unique characteristics, and foremost among them is an external skeleton (technically, not a shell). This exoskeleton protects the animal like a suit of armor while serving as its skeleton. The muscles of arthropods are connected to the inside of the exoskeleton.

The exoskeleton is made of a tough substance called chitin, which is secreted by cells in the epidermis. Because the exoskeleton cannot expand, arthropods periodically shed their armor in order to grow a larger exoskeleton. This process is called molting. The animal must first grow a soft exoskeleton beneath the principal exoskeleton. When the animal grows large enough, it cracks open the old exoskeleton and crawls out, allowing the new, softer

one to grow and harden. New exoskeletons are usually a size or two larger than the arthropod in order to permit further growth and allow for reasonable periods between molts. Immediately after molting, arthropods are vulnerable because they are defenseless. During this time, the animal hides and waits for its armor to harden. Lobsters caught after molting are sometimes called soft-shelled because they feel noticeably soft.

The name *arthropod* means 'jointed-foot.' In order for the arthropod to move in such a rigid body, it has numerous joints in its exoskeleton that flex like door hinges. A lobster is flexible and can rotate its claws enough to pinch an attacker.

Arthropods have an "open" circulatory system. Lacking arteries, veins, and capillaries, blood is pumped through sinuses (open spaces) within the animal in order to reach the tissues.

Most people know that insects have compound eyes. The eyes of marine arthropods are no different. Each eye comprises many smaller, light-sensitive organs, called ommatidia. Together, these ommatidia form a single working eye. The compound eye seems specialized more for detecting motion than for detailed sight.

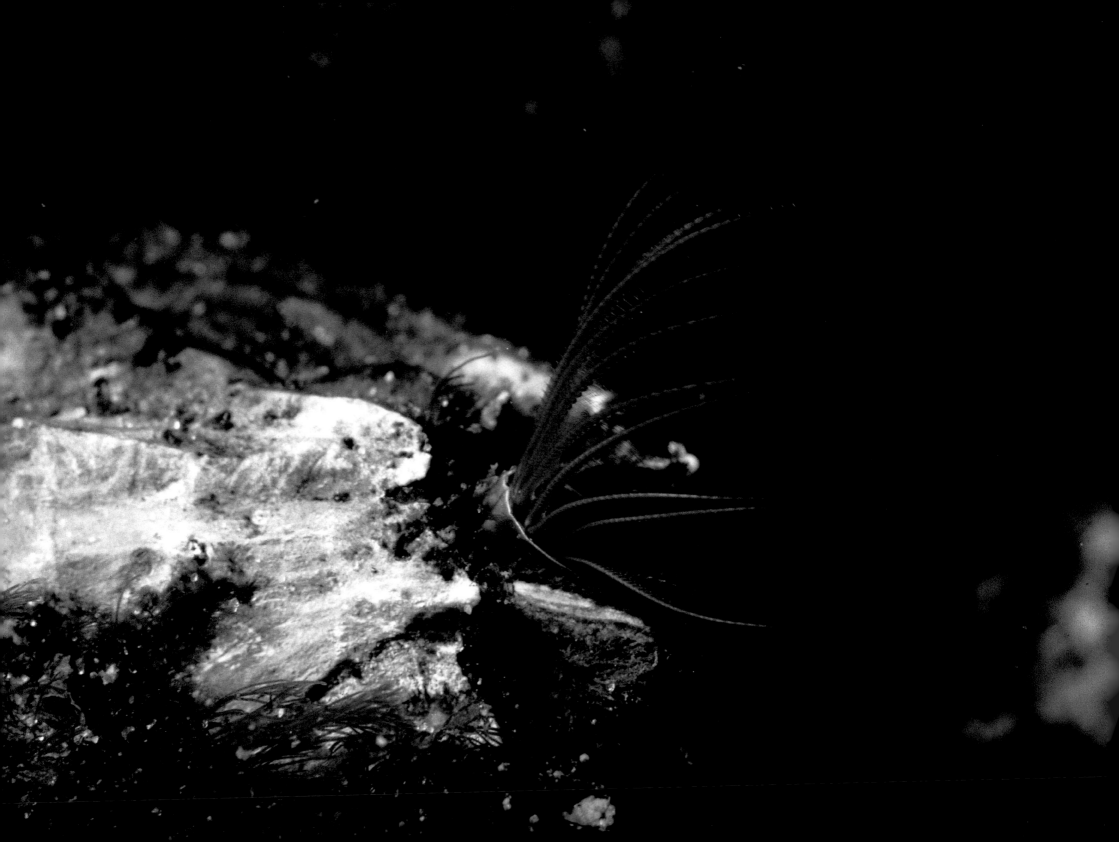

ARTHROPODS

A FEW WORDS ABOUT CHITIN

Every year throughout the world, millions of tons of arthropod armor are thrown away. Since there is no widely accepted use for these exoskeletons, they become useless landfill. This may soon change as scientists begin to invent new uses for these "shells." Chitin is a polysaccharide closely related to cellulose, which is perhaps best known for its use in making paper. Chitin can be chemically transformed into a material called chitosan, from which a number of things can be made. Among the most interesting of these is a type of edible, non-toxic, biodegradable film that can be made into environmentally friendly packaging or even digestible drug capsules. Chitosan can also be made into bandages that keep microorganisms from growing on wounds but are completely nontoxic to humans. Chitosan may also be valuable in wastewater treatment, where it helps organic material clump together so it will settle out of water faster. There are hundreds of possible uses for this abundant resource, but making them a reality will mean collecting all those crab "shells," finding a cost-effective method for creating the chitosan, and finally, getting businesses to use this new product. Many scientists believe, however, that chitin could become as widely used as its close cousin, cellulose.

THE SUBPHYLUM CRUSTACEA

Crustaceans are probably best known as a dinner course. Lobsters, crabs, and shrimp, as well as barnacles, amphipods, isopods, and copepods are all Crustaceans. The subphylum is defined by the five pairs of appendages carried by its members. Usually the front pair, called Chelipeds, have claws, while the remaining four pairs are legs for walking, although in many species, the second and even the third pairs of legs have smaller pincers.

Although some primitive crustaceans have a single body piece, called a trunk, more advanced forms (like lobsters, crabs, and shrimp) have bodies that are divided into two regions: a cephalothorax (or thorax) and an abdomen. The cephalothorax (literally 'head-body') is so called because it contains the "head" as well as the main body organs. The abdomen, which many people erroneously call the tail, is largely muscle. The abdomen is usually made up of six segments, each with a pair of *swimmerets* (small legs). In males, the most forward pair of swimmerets are longer than the others and are designed for inserting sperm into the female. The last segment of the abdomen ends in a flattened section called the telson. The tailfan is composed of the telson and two flat appendages on each side called uropods.

Most crustaceans are similar in appearance, though barnacles are an exception. That barnacles are crustaceans is obvious only during their larval stage. There are about nine hundred different species of barnacles, which are noted for their special glands, which produce a "cement" that allows them to stick to rocks, ships, whales, docks, and just about any other hard surface.

As barnacles grow, they create their own little "houses," called carina, which look like tiny volcanoes. Carinas are made of calcium carbonate, which barnacles manufacture by combining carbon dioxide and calcium extracted from sea water. Young barnacles float freely as planktonic larvae with no protective carina. If they can avoid becoming a meal for a larger animal, they must eventually find a suitable place to settle and grow before building a carina. Barnacles get their food by waving an arrangement of limbs called cirri in the water to catch drifting plankton. The cirri are present, as legs, even on larval barnacles. When the barnacle reaches its adult form, it attaches its head to the substrate and waves its legs to collect plankton.

◄ *Barnacles are basically shrimps which build a limestone house and wave their legs in the water to catch plankton. Inside the house, called a carina, the barnacle is attached by its head to the substrate so it can use its legs, called cirri, for food collection.*

ARTHROPODS

 **THE DECAPODS**

The crustaceans familiar to most people – lobsters, crabs, and shrimp – are part of the order Decapoda (meaning 'ten-legged'), which is itself part of the larger class Malacostraca, which includes nearly three-quarters of all known crustaceans.

The royalty among North Atlantic decapods must be the Northern Lobster (*Homarus americanus*), which grows to 34 inches (86 cm) long and stands 9 inches (23 cm) high. Found from northern Canada to Virginia, lobsters are perhaps best known for their powerful claws. The two dissimilar claws serve different purposes. The larger one, referred to as the crusher, is designed to crack hard objects like snails and clams. The smaller claw, known as the cutter, is used to tear apart the prey on which the animal feeds. When lobsters fight over territory or food they sometimes lose one or both claws in the battle, but in a remarkable evolutionary feat of survival these claws can be regenerated over time.

Estimates based on the size of lobsters versus the rate at which they are known to grow suggest that Northern Lobsters can live for more than one hundred years. In fact, a lobster does not reach legal harvesting size until it is about seven years old.

Although lobster is widely considered an epicurean delicacy, knowing more about the lobster's dietary habits might discourage some people from ordering one at their favorite restaurant. Lobsters, like crabs and shrimp, are mostly scavengers that eat what they find on the ocean floor. Usually, that includes things that have been dead for some time, which explains the attraction of fish heads when used as bait in lobster traps. Lobsters also consume sewage, which explains why they thrive in Boston Harbor. Lobsters are indiscriminate about what they eat, and divers find they can hand-feed them fish scraps. It is an interesting way to get close to a lobster that would not normally consort with a diver. After all, divers frequently catch and eat lobsters.

In Maine one night, fellow diver Tom Krasuski and I encountered a huge lobster foraging for food. Lobsters are nocturnal animals and thus difficult to find during the day but easy to spot at night. I picked up the 20-pound (9 kg) behemouth so that Tom could take a picture. I had to hold it a full arm's distance away from my body to avoid its huge claws, which were as large as my head. When I put the lobster down, it tore off in pursuit of Tom, either out of annoyance or sensing the prospect of a meal. I have never seen a diver move so quickly! Tom stayed off the bottom and kept a wary eye during the balance of the dive.

 **CRABS**

The crab is physically similar to the lobster, but its abdomen is folded out of sight under its thorax. Crabs frequently use this body modification to hold eggs until they hatch. The Atlantic Rock Crab (*Cancer irroratus*) is probably the most familiar crab in the North Atlantic. It is common from the Arctic to South Carolina and may be found in both shallow water and in depths of greater than 2,000 feet (610 m). Though they reach a maximum size of only about $5\frac{1}{2}$ inches (14 cm) wide, they are territorial and can defend themselves fiercely when disturbed.

Rock Crabs display an interesting mating ritual common to many species of arthropods. Crabs usually release a pheromone to attract each other, and Rock Crabs are no exception. (The males of some species of fiddler crabs actually attract a mate by waving their large claw around in a pattern designed to impress the opposite sex.) Because mating cannot take place until the female has molted and has a soft shell, she will wait until a molt is imminent and then seek out a mate. After mating, her partner will protect her by carrying her around with him night and day until her shell has hardened. When her molt hardens, she will leave him.

◄ *The Northern lobster* (**Homarus americanus**) *is a favorite food on the east coast of North America. This Crustacean can probably live to 100 years of age!*

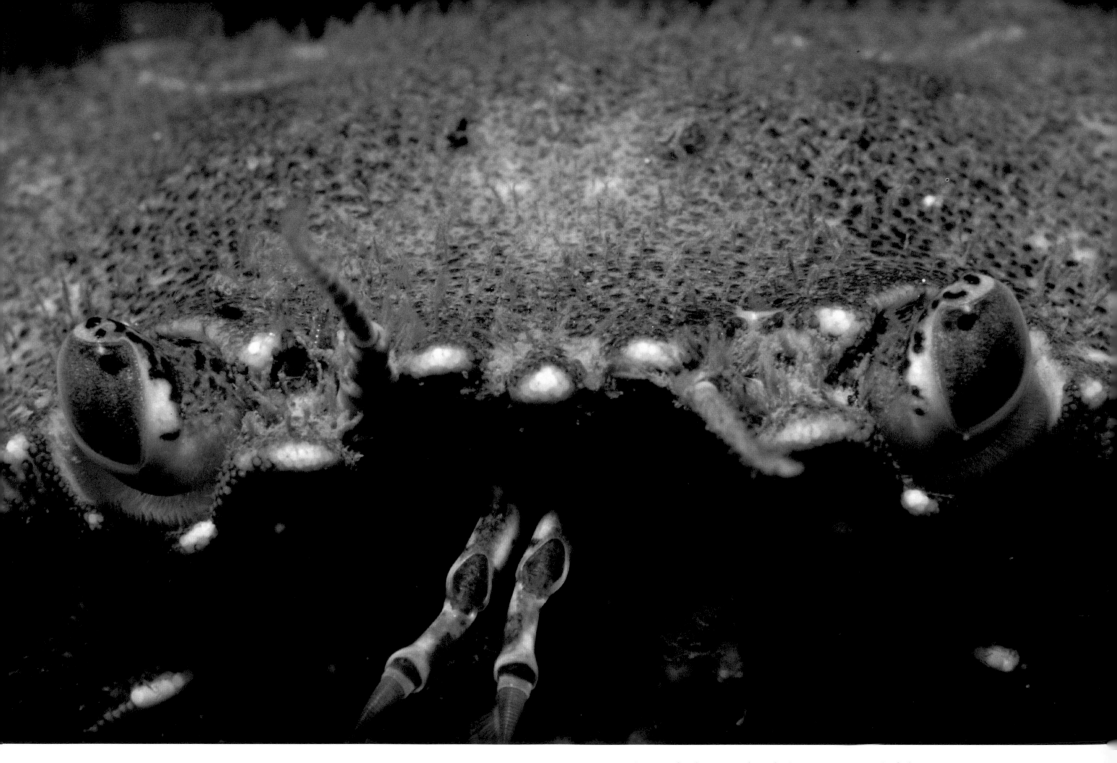

▲ *A face only a mother could love: the Atlantic Rock crab.*

*A pair of Atlantic Rock crabs (**Cancer irroratus**) fighting it out over territory.* ▶

Chapter 5 **ARTHROPODS**

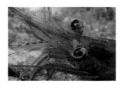 # ARTHROPODS

The Common Spider Crab (*Libinia emarginata*) is easily recognized by its long, skinny legs. Although they grow to only about 4 inches (10 cm) in length, the legs can reach 6 inches (15 cm) in length, making for a frighteningly large creature to encounter on a night dive. The Spider Crab allows algae and other marine growth to accumulate on its shell, which acts as an effective camouflage and sometimes makes the crab hard to find. This crab lives in the cold waters of Nova Scotia all the way to the tropical waters of Florida and the Gulf of Mexico and is found from the low-tide line to depths of about 400 feet (122 m).

The Toad Crab (*Hyas araneas*) is sometimes difficult to spot since it cultivates a nice crop of algae, bryozoans, hydrozoans, sponges, and just about anything else it can on its shell. This camouflage can make the crab difficult to identify as a crab. Without their camouflage, Toad Crabs are red to brown in color and grow to about 4 inches (10 cm) in length. They are found from the Arctic to Rhode Island and in waters as deep as 170 feet (52 m).

Hermit crabs choose discarded snail shells as their homes, which ranks them among the oldest recyclers on the planet. When the Acadian Hermit Crab (*Pagurus acadianus*) outgrows its shell, it goes out and finds a larger one. The shell is its defense, and when the crab is disturbed, it simply retracts into its shell and waits.

Hermit crabs sometimes use a form of foreplay in their mating ritual. To coax a female into mating, a male may hold her shell close to him and rap on it or stroke it. This, presumably, is irresistibly attractive to female hermit crabs. Of course, to mate they must both emerge partially from their protective shells, an action that poses its own risks.

The Acadian Hermit Crab is found from Labrador to Chesapeake Bay, at depths of up to 1,600 feet (488 m).

 ## SHRIMPS

Despite their many differences, shrimps, in general, look just like miniature lobsters. Shrimps are a favorite subject for underwater photographers because they are frequently colorful. Although tropical shrimps, like the Cleaner shrimp and Coral shrimp, are noted for their bright coloration, there are several North Atlantic shrimps that easily rival their tropical brethren in beauty.

Montague's Shrimp (*Pandalus montagui*) grows to about 4 inches (10 cm) long and is predominantly red, with tiny patches of bright blue. It is a photographer's dream just to have it stay still. Montague's shrimps are found from the Arctic to Rhode Island but are most common in the northern part of their range – northern Maine and New Brunswick – and live from shallow depths to about 100 feet (30 m). They are also common in Europe, where they are harvested commercially.

The Sand Shrimp (*Crangon septemspinosa*) is almost impossible to find over a sandy bottom because it has evolved to match the color of the sand. Growing to about 3 inches (7$^1/_2$ cm) long, the Sand Shrimp can tolerate large variations in salinity, which may be an adaptation related to its habitat of shallow, sandy bays that receive frequent freshwater run-off. Sand Shrimps are found from the Arctic to Florida in water from only a few feet deep (1 m) to about 300 feet (91$^1/_2$ m) deep.

The Greenland Shrimp (*Lebbeus groenlanicus*) is a small (2-in./5-cm) cold water shrimp found from the Arctic to as far south as Massachusetts. Reddish-brown in color, it is shaped rather like a U, with a sharp downward bend in its abdomen. Greenland Shrimps are found in water from shallow depths to more than 600 feet (183 m) deep and are common in Passamaquoddy Bay.

◄ *The Toad crab (***Hyas araneus***) camouflages itself by cultivating sponges, algae, barnacles, bryozoans and other marine life on its shell. When completely covered, Toad crabs are very hard to spot.*

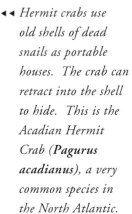

◄◄ *Hermit crabs use old shells of dead snails as portable houses. The crab can retract into the shell to hide. This is the Acadian Hermit Crab (**Pagurus acadianus**), a very common species in the North Atlantic.*

◄ *Here a hermit crab is seen without a shell. When the hermit crab outgrows its shell, it will "try on" new ones as they are encountered. If the fit is good, the crab keeps the new shell and discards the old one. Note the spiral shape of the crab's abdomen, which is an adaptation to living in spiral shells.*

63

◄ *In this close-up shot of a Montague's shrimp, the compound eyes common to all arthropods are easily seen.*

▲ *A close-up of the abdomen of a Montague's Shrimp* (**Pandalus montagui**)*, showing that this female is carrying a bunch of eggs.*

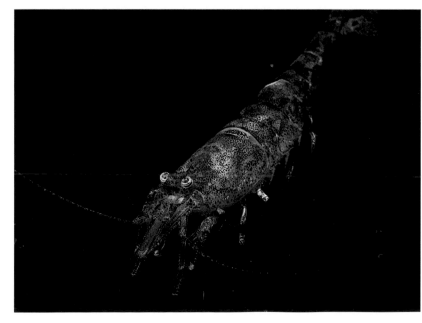

◄ *The brightly colored Montague's shrimp* (**Pandalus montagui**) *rivals tropical shrimp in beauty. This shrimp is about three inches long.*

◄ *The Sand shrimp (**Crangon septemspinosa**) is colored to blend in with sand, but is easily seen against a contrasting background.*

ARTHROPODS

THE ISOPODS

The order Isopoda is familiar to many people because it includes the "pill bugs," which are found under rocks and logs in the forest. These flattened, oval, plate-covered crustaceans are called isopods because their legs are all of approximately the same length. There are myriad isopod species in the ocean. They range in length from less than $1/2$ inch (1cm) to several inches and resemble their terrestrial relatives. Many can actually survive out of water for long periods.

 ## THE AMPHIPODS

While isopods are flattened and oval-shaped, members of the order Amphipoda are usually compressed from side to side. As with isopods, there are both land and marine amphipods. Many people have probably seen the $1/2$-inch-long "beach fleas," that live in sand and washed-up seaweed. Amphipods are a vital link in the food chain and are consumed by corals, anemones, small fish, and other, larger, crustaceans. One of the strangest looking amphipods is the skeleton shrimp, which is not actually a shrimp at all. It has a long, tubular body, with two pairs of long, thin legs at its front and three additional pairs of legs at the rear. They cling to seaweed and other materials, crawling, like inch worms, in search of food.

The whale louse looks similar to the skeleton shrimp but lives as a parasite. After attaching itself to a whale using its hook-ended legs, the louse gnaws a hole in the whale's skin and settles in, away from the current of water rushing past as the whale swims. The louse eats the whale's skin and is a pest, much as a tick is to a dog.

 ## THE CLASS MEROSTOMATA

Related directly to some of the oldest creatures on earth, the members of the Merostomata class are descended from the now-extinct Trilobites, whose fossil remains date them to more than 175 million years ago. There are only five species in the class Merostomata. One, the Horseshoe crab (*Limulus polyphemus*), is found on the east coast of the United States, from the Gulf of Maine to the Gulf of Mexico and is common on Cape Cod. The animal is a tannish-brown color but may accumulate algae growth and thus look green in spots. The other four species are found in Japan, Korea, the East Indies, and the Philippines.

This creature, though technically not a crab and more closely related to spiders and scorpions, has no similar-looking relatives in the North Atlantic. In about 1870, the horseshoe crab was given the name 'horse-foot crab.' Although this name more aptly described its shape, common parlance corrupted the name to the present-day "horseshoe." It has a rounded (horse-foot-shaped) carapace, with a triangular abdomen and a long, slender tail (the telson). On Sir Walter Raleigh's expedition to the New World in 1584, naturalists Thomas Heriot and John White noted that Native Americans used the horseshoe crab's tail (connected to a reed or stick) as a spear tip to spear fish.

The creature's mouth is on its underside and is surrounded by five pairs of walking legs. Each leg is heavily armed with spines on the inside edge of the largest segment. These spines, called gnathobases, grind up food (usually worms, clams, and other small invertebrates) before it is eaten.

Horseshoe crabs prefer shallow water and are rarely found deeper than 50-75 feet (15-23 m). In fact, they lay their eggs at high tide in sand above

◄ *Resembling a tiny dragon, the Polar shrimp (**Lebbeus polaris**) reaches only two inches in length and is found only in very cold water north of Massachusetts.*

A marine amphipod. ▶

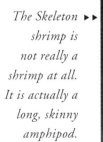

The Skeleton ▶▶
shrimp is
not really a
shrimp at all.
It is actually a
long, skinny
amphipod.

ARTHROPODS

the low-tide line. The 2,000-30,000 eggs in a nest incubate for several weeks while the tide goes in and out. Although the sand keeps the eggs moist, they are out of the water half the time. Horseshoe crabs reach sexual maturity between ages nine and twelve, and their life span may be greater than twenty years.

Horseshoe crab blood, like that of many arthropods, contains hemocyanin to transport oxygen to the cells. Humans have red blood, containing hemoglobin, which is iron-based; hemocyanin is copper-based and makes the blood appear blue instead of red. Because the horseshoe crab has a large ratio of blood relative to its body size compared with other arthropods whose blood contains hemocyanin, it is used by scientists to study hemocyanin.

Why study the blood of an obscure sea creature? *Limulus polyphemus* is naturally immune to endotoxins, germs that are found everywhere. Humans are immune also, as long as the endotoxins do not get into the bloodstream. If these endotoxins do get into human blood, the result is septic shock, which is incurable and usually fatal. Since endotoxins cause about 20 percent of all hospital deaths, its cure would save many lives. Today an extract from

the horseshoe crab's blood, called Limulus Amebocyte Lysate (LAL), is used as an indicator to show the presence of endotoxins in humans. Scientists hope to find a cure for septic shock based on LAL.

More than a matter of merely preserving biological specimens, saving the ocean's creatures may lead to medical advances that will save human lives in the future.

THE CLASS PYCNOGONIDA

Members of the class Pycnogonida are called sea spiders. Although they look like land spiders, they are not closely related. Pycnogonids have eight legs and a very small body. In fact, the body is so compact that the sex organs and part of the digestive tract are located in the legs. These strange arthropods feed mostly on cnidarians (like anemones and hydroids) by inserting a proboscis into the host and sucking out their semiliquid body fluids.

In most species of pycnogonids, the male carries fertilized eggs because he is equipped with

special egg-carrying legs underneath his normal legs. In a few species the female has these legs as well.

Although the sea spiders found in most shallow waters are less than 1 inch (2.5 cm) across, huge, 24-inch (60 cm) specimens have been found in the deep ocean. Various deep-water pycnogonids have been found as deep as 12,000 feet (3,658 m). Although pycnogonids are found in all the oceans, their numbers are most concentrated in the Arctic and Antarctic waters.

◄ *Although the Horseshoe crab (**Limulus polyphemus**) isn't really a crab, it is sometimes referred to as a "living fossil" because it has ancestors dating back 175 million years!*

◄◄ *The bases of the walking legs of the Horseshoe crab are covered with spines called Gnathobases. These spines allow the Horseshoe crab to grind food up before eating it.*

◄ *Reaching only ¹/₂" in length, the Anemone sea spider (**Pycnogonum littorale**) drinks the body fluids of anemones by sucking it out with a probiscus. In spite of its name, the sea spider is not closely related to land spiders.*

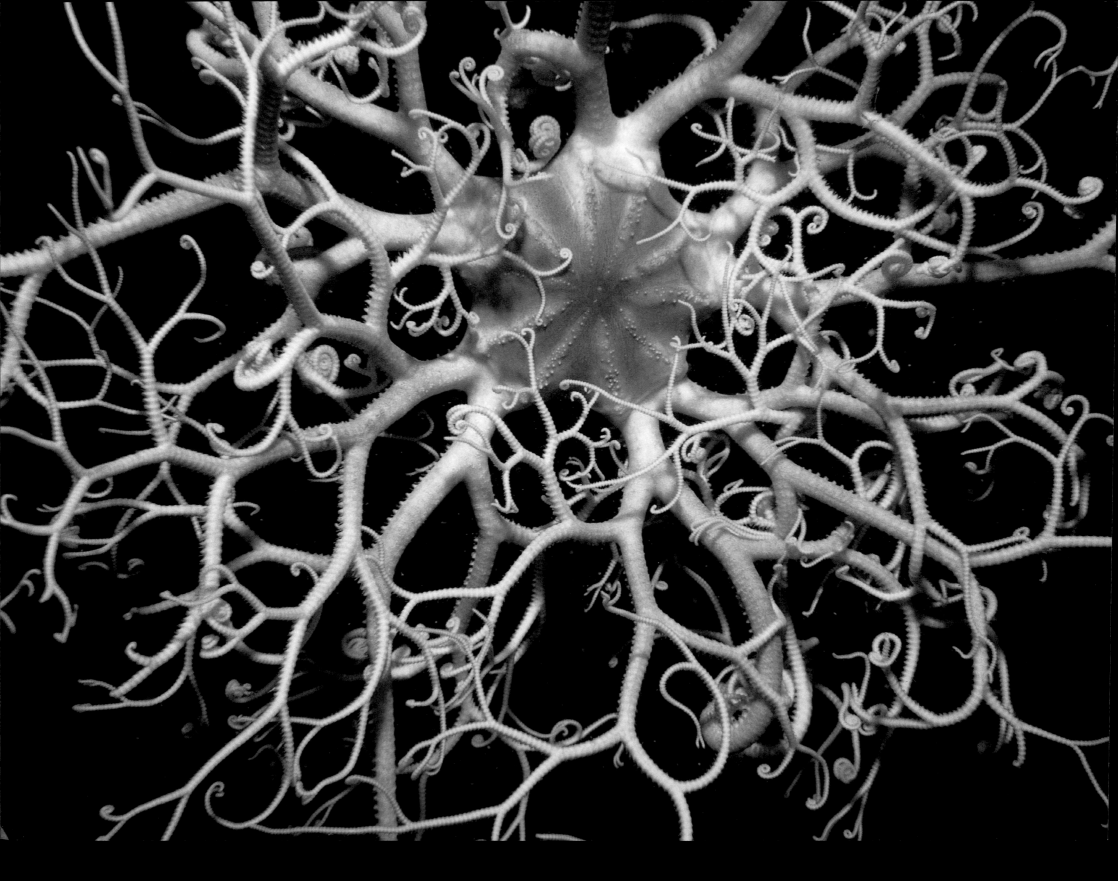

ECHINODERMS

The six thousand species of the phylum Echinodermata live only in the ocean. The phylum's name, from Greek, means 'spiny skin,' and although many echinoderms actually do have spiny skin, others do not. All echinoderms do have in common something called radial symmetry. That means their appendages (arms or *rays*) point outward from the center of their bodies like the spokes on a bicycle wheel. The appendages usually occur in multiples of five, although there are a few exceptions. Several members of this group – sea stars and sea urchins – are familiar, and their radial symmetry is obvious.

Less obvious is the water vascular system common to all echinoderms. Examining the oral (ventral) surface of a sea star reveals hundreds of tiny feet usually arranged into several rows on each ray of the star. In most echinoderms these tube feet, or podia, are filled with sea water. The water vascular system within the body of the animal is also filled with sea water. By expanding and contracting chambers within the water vascular system, the echinoderm can force water into its tube feet to extend them. Muscles in the tube feet retract them. By expanding and retracting its tube feet, the

creature can walk. Many echinoderms also have suckers at the ends of their tube feet, which can be used to capture and hold prey or to hold on to rocks in a swift current or tide.

Interestingly, although most mature echinoderms are benthic, the larvae are usually planktonic with bilateral symmetry. As they mature, echinoderms change their body shape from bilaterally symmetrical to radially symmetrical and then settle down to the sea floor.

■ THE STELLEROIDS

Perhaps the most familiar echinoderm is the sea star. Although commonly called a star fish, sea stars are not fish at all. Scientists reserve the term *fish* for vertebrates with fins; sea stars have neither vertebrae nor fins.

The subphylum Stelleroidea contains the two classes of sea stars: the class Asteroidea contains the true sea stars and sun stars; the class Ophiuroidea contains the brittle stars and basket

 # ECHINODERMS

stars. The distinction between the two classes lies in the way the arms connect with the body. Ophiuroids have a distinct central body part (called a central disk) with arms radiating out. Adjacent arms do not connect with one another. Asteroids, on the other hand, have arms that appear to connect to one another, rather than to a central disk.

The sea star's aboral (top) surface looks spiny on close examination. Its rumpled skin contains several different types of formations. Some of the bumps on the surface, called dermal branchiae, absorb oxygen from the water. Another adaptation, pincerlike pairs of organs called pedicellaria, pluck things from the sea star's skin that are a nuisance. For example, the larval form of a barnacle could settle onto a sea star and begin to grow if the sea star had no way to remove it.

There is a single-colored spot on the aboral surface of sea stars, which is called the madreporite. This calcareous piece of the water vascular system is perforated with tiny holes and serves as the filter between the animal's water vascular system and the ocean.

Sea stars have a light sensitive organ at the tip of each ray, called an eyespot. When moving across the ocean floor, the sea star usually leads with one ray, probing the surface ahead. Although the star cannot see in the same way we do, it can detect the presence and direction of light and does seem to have a sense of direction.

◄ *A close up of the aboral (top) surface of the Northern sea star (**Asterias vulgaris**) reveals the characteristic "spiny" echinoderm skin.*

Sea stars are capable of regenerating limbs when one is severed or damaged. In a few species, the severed limb grows into a new sea star, but in most cases, the severed limb dies.

Sea stars eat a variety of marine creatures, including barnacles, clams, mussels, snails, sea urchins, and sometimes even other sea stars. Many sea stars, such as the Northern Sea Star, eat mussels and clams in a fascinating way. First, the sea star wraps its rays around the shell of the intended victim. Then it applies suction with its tube feet to pull the two mussel shells apart. The sea star does not need to apply force for long to tire out the mussel. The sea star can apply so much force to the mussel's valves (7 lbs, or 3 kg, or more) that it actually bends the shell. Seizing the moment, the sea star then extends its stomach through its mouth into the mussel. Remarkably, this requires an opening of only 1/250th of an inch (0.1 mm). The shell becomes a lunch box, and when the star is finished with the mussel, only a shell remains.

 ## THE TALE OF THE HUNGRY SEA STARS

Despite their benign appearance, sea stars are voracious eaters. To film the tube feet of a sea star in action for Oceanic Research Group's first educational film, we organized a marine aquarium. We placed two small (3-inch/7-cm) Northern Sea Stars, both of which were missing a limb, three small urchins, eight mussels, two limpets, five periwinkles, a large rock

covered with barnacles, and a variety of fish, crabs, and anemones into a 55-gallon aquarium. Within two weeks, the sea stars had eaten every urchin, mussel, limpet, periwinkle, and barnacle. They could not eat the fish, crabs, or anemones, but they probably tried. After filming, we were so interested in these animals that we decided to keep the tank going to see how long it would take for the sea stars to regenerate their limbs.

Since feeding sea stars a live diet was simply too much work, we experimented to see whether sea stars could learn to eat tropical fish food pellets. The pellets floated for several hours before sinking to the bottom, where the sea stars found and ate them. After several weeks, however, the sea stars discovered that the pellets floated. They learned to make their way up the tank walls at feeding time to capture pellets floating on the surface. This behavior required considerable effort on the sea stars' part and involved bending a limb back onto the water's surface to capture floating pellets. The sea stars demonstrated great adaptability in working out a totally new and unnatural way to eat. But there is more to this story.

After a few more weeks, the response time of the sea stars measured less than 15 seconds between the instant when the pellets were poured into the tank and when the echinoderms first moved toward the surface. Initially, it seemed that the sea stars had a fairly sophisticated form of "smell," but perhaps not.

In another few weeks, simply opening the lid of the tank prompted the sea stars to move surfaceward. Whether because the sea star's "primi-

 # ECHINODERMS

▼ *The Northern sea star (Asterias vulgaris) engaging in a meal of mussel. The sea star will force the mussel open and inject its stomach into the mussel to partially digest its victim outside of its body.*

tive" sight is better than we believe or because they were able to sense the vibrations of opening the tank lid, clearly sea stars can learn and quickly adapt to changing environments. Incidentally, it took about a year for the animals to regenerate their lost limbs.

 ## REPRODUCTION BY FISSION

Many sea stars use their regenerative abilities for reproduction. Several species of *fissiparous* sea stars reproduce by fission. In this process the sea star literally tears itself in two, leaving a gaping wound in each remaining half. Soon, the wound heals, and each half of the sea star goes on with life as a normal animal, slowly regenerating its lost limbs. This asexual reproductive technique has advantages and disadvantages.

Sexual reproduction involves the release of planktonic larvae to drift with the currents and settle elsewhere. In areas where strong currents sweep the larvae into deep water, juvenile sea stars may never mature. Furthermore, many planktonic larvae are consumed by planktivorous creatures, putting the odds of survival to maturity at less than one in a million. Asexual reproduction by fission may be how sea stars try to beat the odds. By splitting in two, the sea star can be reasonably sure to produce an offspring that will survive. On the other

hand, fission is a process that requires much energy and can have disastrous consequences if the sea star is attacked during, or just after, fission takes place.

There are several species of tropical sea stars of the genus *Linckia* that have taken this process to another level. In these echinoderms, reproduction takes place by autotomy, or voluntary dismemberment, instead of by fission. The sea star periodically sheds a single limb, which can regrow an entire body. As the wound heals in the parent sea star and the missing limb begins to regenerate, another limb is shed. This continues until each limb has been shed once. The parent then retires from reproduction until all its limbs are fully regenerated. Then, it may start all over again. Meanwhile, each autotomized limb grows into a fully developed sea star and begins the same process.

Reproduction by autotomy guarantees a reasonably good survival rate among offspring, thereby eliminating the risks and uncertainty involved in planktonic development. For every problem, there is more than one solution, and these sea stars have taken a novel approach to overcoming the poor odds of planktonic development.

 ## TYPES OF SEA STARS IN THE NORTH ATLANTIC

The North Atlantic is home to a variety of sea stars, including but not limited to: the Northern Sea Star, Blood Star, Spiny Sun Star, Basket Star, Horse Star, and the Brittle Star.

◀ *The Northern sea star (Asterias vulgaris) on the right is known as a "comet." It is the autotomized (dismembered) arm of a sea star regenerating a new body and arms.*

◄ *A Blood star (**Henricia sanguinolenta**) crawling across a rock covered with*
a purple coraline algae called Lithothamnion. This sea star can absorb
nutrients directly through its skin, but it also eats sponges.

▲ *A 12 armed Spiny sun star (**Crossaster papposus**) moving*
across a rock in search of a meal. These large sun stars can have
between 8 and 14 arms and prefer to eat other sea stars!

▲ *A beautiful Smooth sun star (**Solaster endeca**) photographed in Maine.*

Unlike most sea stars which produce planktonic larvae, the Winged sea star ▶
*(**Pteraster militaris**) broods its young in pouches on its underside.*

Chapter 6 **ECHINODERMS**

◄ *In order to capture prey, the mature Northern basket star* (**Gorgonocephalus articus**) *raises several of its arms into the water current to capture drifting plankton.*

The wide-ranging Northern Sea Star (*Asterias vulgaris*) lives in waters from Labrador to North Carolina. It is found from the middle-tide zone to as deep as 1,100 feet (335 m). The Northern Sea Star is easily the most common star throughout New England. It has five appendages and comes in many colors, including pink, rose, orange, tan, cream, gray, a greenish tint, a bluish tint, lavender, and light purple. Each arm tip has a red eyespot, and there are four rows of tube feet in each groove on each appendage. The star can grow to 16 inches (40 cm) and has suckers on its tube feet to hold prey readily. The grip of a Northern Sea Star on a mussel is so tenacious that attempting to separate it from a mussel will often result in the loss of a number of its tube feet. An autotomized limb of this species can regenerate a completely new sea star as long as the limb still has a small section of the central disk attached. Studies have shown that if a severed limb has at least one-fifth of the central disk attached, a new star will grow. Although there is no research that shows that the Northern Sea Star reproduces using this technique, at certain times of the year large numbers of *Asterias vulgaris* are seen to be missing two to three limbs. It seems plausible, therefore, that the Northern Sea Star can reproduce by fission.

Forbes' Common Sea Star (*Asterias forbesi*) is similar to the Northern Sea Star in general shape and range. It prefers somewhat warmer

The juvenile Northern basket Star ▶
(**Gorgonocephalus articus**) *eats soft coral by wrapping itself around the coral and chewing away on the polyps.*

▼ *The elusive Daisy brittle star (**Ophiopholis aculeata**) is seen here at night crawling across a Smooth sun star (**Solaster endeca**) in an effort to escape the photographer's camera. Brittle stars get their name from the fact that they are extremely fragile, but they can move surprisingly quickly.*

◄ *When the current gets too strong or predators close in, the basket star can bundle itself up and hang on for dear life!*

water and is found from the Gulf of Maine to the waters of Texas. This sea star reaches about 10 inches (25 cm) across and eats bivalve mollusks. A small specimen can exert a 12-pound (5.5 kg) pull on the bivalve's shells and requires only a minuscule gap between the shells to insert its stomach into the victim. Although this sea star is found in many color schemes (but usually a brownish color or tan), it always has an orange madreporite.

The Blood Star (*Henricia sanguinolenta*) is named for the blood-red color typical of its upper side, although it can also be found in pink, a purplish color, orange, yellow, cream, and white. Blood Stars have five rays, can grow to 8 inches (20 cm) across, and are found from below the low-tide line to depths of almost 8,000 feet (2,440 m). They range from the Arctic to Cape Hatteras, North Carolina. The Blood Star eats by trapping organic particles in its mucus and then moving the particles to its mouth by ciliary action. It eats sponges and also absorbs nutrients from the water directly through its skin. Although the skin of the Blood Star is much smoother than that of the Northern Sea Star, magnification reveals that the skin has the bumpy surface common to most echinoderms. Unlike many other sea stars, the Blood Star broods its young from an egg to its benthic form, skipping the planktonic larval stage.

Although the Spiny Sun Star (*Crossaster papposus*) grows to only 7 inches (18 cm), it is one of the more beautiful and colorful echinoderms. This star can have between 8 and 14 arms, although 10-12 seems most common. The star is usually red, with concentric rings of pink, yellow, white, or dark red. It is found from the Arctic to the Gulf of Maine and from the low-tide line to 1,000 feet (305 m) deep. On a trip to northern Maine, I discovered a sun star feeding on a young blood star. Sun stars prey almost exclusively on other sea stars and do not bother

chewing their food; they swallow it whole. There are no table manners in the world of the sea stars.

The Smooth Sun Star (*Solaster endeca*) is well represented in the Gulf of Maine, reaching 16 inches (40 cm) across. It is not found south of Cape Cod. This brilliantly colored sun star can be purple, red, orange, or pink and can have between 7 and 14 arms. It is frequently found near large numbers of sea cucumbers, which are a favorite food. It is one of a small number of sea stars that have no planktonic stage and develop directly from an egg to a benthic creature.

The Horse Star (*Hippasteria phrygiana*) is a five-armed (and nearly pentagonal) cold-water star that grows to about 8 inches (20 cm) and is found as deep as 2,600 feet (793 m). Unlike the Northern Sea Star or Spiny Sun Star, this creature has no preferred diet and eats almost anything it can find, from mussels, worms, and urchins to anemones, soft corals, and other sea stars. Its range stretches from the Arctic to Cape Cod.

The Winged Sea Star (*Pteraster militaris*) is a thick-bodied, five-armed sea star that reaches about 51/2 inches (14 cm) across. It is usually red to yellow on the aboral surface, and white or tan underneath. It is found from the Arctic to Cape Cod in water from a few feet deep (1 m) to more than 3,000 feet (915 m). Its reproductive habit is particularly unusual. Unlike most sea stars, which spawn planktonic larvae, the Winged Sea Star has a "pouch" covered by a membrane that is used for brooding its young. Eggs are deposited in the pouch, where they hatch into tiny sea stars. They

*The mouth of the Green sea ▶ urchin (**Strongylocentrotu droebachiensis**), found on its underside. The detail of the 5 teeth, called "Aristotle's Lantern" is visible.*

 # ECHINODERMS

then break through the membrane and walk away as tiny replicas of their parents. There is no planktonic stage of development.

The Northern Basket Star (*Gorgonocephalus articus*) is distinct from the other stars found in the North Atlantic. It is an Ophiuroid and has a distinguishable central disk body, with five arms. The arms branch apart several times as they get farther from the body. The arms farthest from the body are very small and numerous. The mouth on the underside has five jaws, and the creature has tiny tube feet. Its color is yellowish-brown to dark brown, and the arms are a yellowish-tan to white. The Northern Basket Star moves with surprising speed and can coil itself around objects into what seems an almost impenetrable Gordian knot.

Juveniles eat soft coral (*Gersemia rubiformis*) by wrapping themselves around a colony and gnawing away at the polyps. Adults use their branched arms to collect plankton as it drifts by in the current.

Although Northern Basket Stars reportedly range from the Arctic as far south as Cape Cod, I have never seen one south of northern Maine. They live in water between the low-tide line to depths of almost 5,000 feet (1,524 m).

Another Ophiuroid, the Daisy Brittle Star (*Ophiopholis aculeata*), is fairly common in the North Atlantic but is rarely seen because it tends to hide in cracks and holes. I once brought a medium-sized rock home from the ocean in a bucket because the rock had a nice sample of Lithothamnion (a pretty purple coraline algae) growing on it. Days after the rock was put into our research tank, I noticed a tiny brittle star ray sticking out of a crack in the rock. I had unwittingly moved the home of the brittle star along with its inhabitant. The Daisy Brittle Star has five arms and grows to about 8 inches (20 cm). It is found from the low-tide line to water more than 5,000 feet deep, from the Arctic to Cape Cod.

Brittle stars are nocturnal and easier to find at night when they are prowling for food. In the daytime, an exposed ray or two often reveals their hiding places in the rocks. Removing the rocks may uncover the entire animal, but beware: the brittle star moves like no other sea star. It can scramble away from trouble with surprising speed.

Why are they called brittle stars? Their rays are extremely fragile. Any attempt to pull a brittle star out of its hole by one of its rays, no matter how gentle the attempt, will result in the loss of a limb. They can regenerate their rays in about a year. The tube feet on the rays transport food particles from the arm spines to the mouth.

 ## ECHINOIDS

The class Echinoidea includes sea urchins, heart urchins, cake urchins, and sand dollars. A sea urchin's body is covered with sharp spines, which offer protection from many would-be predators. The spines are joined to the skeleton of the animal, called the test, with a structure like an automobile ball joint. Muscles attached to each spine enable the urchin to swivel them in the direction of a predator. The test is an egg-shaped spherical structure constructed of rows of radially arranged plates fused together.

The creature has five paired rows of tube feet that carry suckers and can be extended past the long spines. The anus is on the top of the creature, and the mouth is on the underside. The mouth contains five teeth that point toward the center of the mouth. This structure looks and works like the jaws of a drill chuck and is called Aristotle's Lantern because it was first described as looking like the top of an oil lamp in a book by Aristotle. The creature uses its tube feet to pull itself against the substrate so it can gnaw away at algae with its mouth.

The Green Sea Urchin (*Strongylocentrotus droebachiensis*) is the only species that is plentiful in New England. It occurs from the Arctic to New Jersey in a wide range of depths, from the low-tide line to as deep as 4,000 feet (305 m). They grow to about 3 inches ($7^1/_2$ cm) across and feed primarily on algae and decomposing matter but are indiscriminate eaters. Green Sea Urchins are so abundant in New England that they carpet the floor of many ocean coves.

Sushi (raw fish) is not one of my favorite dishes, but the Japanese consider it a delicacy. Among sushi dishes, *uni* (raw sea urchin) is one of the most sought after. Not all urchins are edible, but the Green Sea Urchin is, and New England supplies much of the Japanese urchin market. The gonads of the urchin are the only part of the animal that is eaten. (A friend who enjoys sea urchin persuaded me to try it. The gonads look like yellow garden

 # ECHINODERMS

slugs and have the consistency of soggy bread. They taste like moldy cheese and have a strong aftertaste.)

At a salmon farm on an ocean inlet in Eastport, Maine, the fish are raised in floating pens. Diving under the pens revealed an ocean floor blanketed with Green Sea Urchins. Thousands of them live well on fish waste and a steady supply of food that regularly slips past the fish at feeding time.

The urchins keep the bottom clean under the salmon pens and prevent possible disease by consuming uneaten food and waste before it decomposes. Not long after my visit, the farm owner decided to try harvesting urchins. He built a small hand-held, rakelike device that he pulled along the bottom. One harvest netted several thousand dollars in urchins, but he was careful to harvest only a portion of the total population of urchins in order to leave plenty of "vacuum cleaners" under the farm.

Historically there has been no local demand for sea urchins, but recent interest from the Japanese prompted arguments for a sea urchin "fishing season" in order to keep the Green Sea Urchin from being overfished. Today, the states of Maine, New Hampshire, and Massachusetts all have urchin fishing seasons and minimum-size requirement laws.

Many sandy areas of ocean floor are carpeted with sand dollars. A living sand dollar, however, does not look like the bleached white test that is found on the shore. Sand dollars, heart urchins, and cake urchins all differ from sea urchins in that they have evolved several modifications that allow them

to burrow into the ocean floor. The spines are still present, but they are much shorter (most notably in the case of the sand dollar) and more numerous.

In the North Atlantic, the Common Sand Dollar (*Echinarachnius parma*) predominates. Its range extends from Labrador to Maryland, where it lives from the low-tide line to depths of 5,000 feet (1,525 m). The sand dollar eats fine particles of organic matter it extracts from the sand.

HOLOTHUROIDS

The class Holothuroidea includes creatures called sea cucumbers, so named because they resemble the garden cucumber. The similarity ends there. Resting on the ocean floor like footballs, sea cucumbers have five rows of tube feet running lengthwise, like the seams of a football. Three of the rows of tube feet are well developed and are in contact with the substrate. Two rows are usually underdeveloped or are missing entirely and are not used. The mouth at one end is surrounded by tentacles. These tentacles, usually branched, are actually specialized tube feet and are part of the water vascular system. Unlike sea stars and urchins, however, the sea cucumber's vascular system uses its own body fluid, so there is no direct interface (madreporite) between ocean water and the internal organs of the animal.

The sea cucumber feeds in a fascinating way. It positions itself in a spot on the ocean floor where a current will bring a steady supply of food (plankton and other organic particles) its way. The tentacles spread open to collect food. Then, the cucumber inserts one tentacle at a time into its mouth, cleans it off, and holds it out to collect more food while it cleans the next tentacle. The sea cucumber does this for hours at a time.

The body fluid of many sea cucumbers is poisonous. (In an aquarium, if an injured sea cucumber releases this fluid it can kill the fish.) The poison of some sea cucumbers has shown promise as the source for a drug that can inhibit the growth of cancer cells.

Some sea cucumbers employ an unusual method of self-defense. When a sea cucumber is attacked, it can expel some of its internal organs. The effect either satisfies a predator or scares it off. The cucumber then grows another set of organs.

Some sea cucumbers secrete a sticky substance as a self-defense measure. This gluelike substance is so sticky that it cannot be removed from human skin without tearing off the hair it touches. Historically, people have used this substance as a bandage to bind wounds.

The Chinese eat certain sea cucumbers and consider them delicious. Some species have edible muscular body walls. The texture has been described as gelatinous.

◄ *A juvenile Orange Footed sea cucumber (**Cucumaria frondosa**) rests among its relatives, the Green sea urchins.*

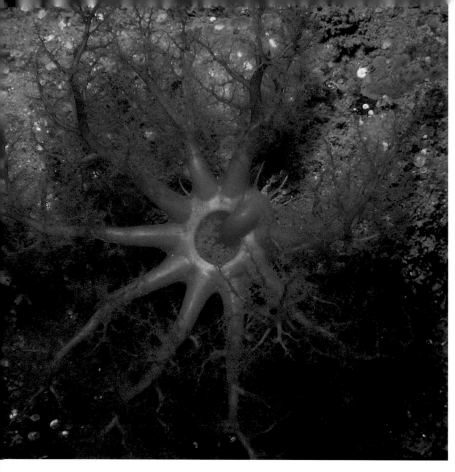

▲ *As seen from the top, the Scarlet Psolus uses its ten tentacles to capture planktonic prey. The tentacles are part of the sea cucumber's water vascular system, controlled by means of internal water pressure like a hydraulic system.*

*Side view of the Scarlet Psolus (**Psolus fabricii**) with it tentacles extended. This striking red sea cucumber, unlike most sea cucumbers, has a hard body made of calcareous plates.* ▶

◄ *The mouth and tentacles of a feeding Orange Footed sea cucumber (**Cucumaria frondosa**). The arms are held out into the passing current to collect plankton, and then the sea cucumber licks off each arm, one by one.*

develop into planktonic larvae, which eventually settle down to the sea floor for their benthic life.

The body surface of most sea cucumbers is leathery in appearance and feel, but, as always, there are exceptions. One is the Scarlet Psolus (*Psolus fabricii*). This holothuroid hardly resembles the Orange-Footed Sea Cucumber. It has a small body that attaches firmly to rocks. When retracted, it resembles a tiny volcano. The Scarlet Psolus is one of a small number of cucumbers with a body made of calcareous plates. It does not, therefore, have the appearance or feel of leather, but rather more of rock or shell. When the creature is extended, the bright red tentacles are striking and work like those of the Orange-Footed Cucumber to collect food.

The first time I saw a Scarlet Psolus, I thought it was soft coral until, as I approached, it retracted out of sight. The branching tentacles of the Scarlet Psolus are remarkably similar, both in color and in shape, to branching soft coral.

It is difficult to photograph these creatures because they are so sensitive to motion in the water. It is often impossible to get within 2 feet ($^2/_3$ m) of the animal before it senses danger and closes up. The trick to approaching the Scarlet Psolus undetected is to use a strong current as a cover. The current seems to obscure the turbulence created by a diver and keeps the Psolus unaware of an alien presence. The Scarlet Psolus is found from the Arctic to Cape Cod from the low-tide line to depths greater than 1,300 feet (396 m).

SEA CUCUMBERS OF THE NORTH ATLANTIC

The Orange-Footed Sea Cucumber (*Cucumaria frondosa*) has orange-colored tube feet that give it its name. The animal is a brownish to greenish color, and the tentacles can range from dark brown to pure white. This is the largest and most common sea cucumber in New England; it is found from the Arctic to Cape Cod, especially in areas that have a swiftly flowing current, like the Bay of Fundy. This cucumber grows to a massive 19 inches ($48^1/_4$ cm) long and is found as deep as 1,200 feet (366 m). It has ten highly branched tentacles.

Watching these creatures eat is fascinating, but getting too close will cause the cucumber to hide by retracting all its tentacles and closing up. When it does this, it looks like a football.

A dive in Bar Harbor, Maine, near Acadia National Park, revealed hundreds of these cucumbers, which appeared to be piled together on rocks beneath the surface. It looked like a locker room after football practice. Sea cucumbers have separate sexes and reproduce sexually. Although some sea cucumbers brood their eggs, Orange-Footed Sea Cucumbers merely release eggs and sperm into the water for fertilization, which requires perfect timing between the males and females. The fertilized eggs

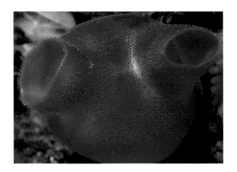

UROCHORDATES

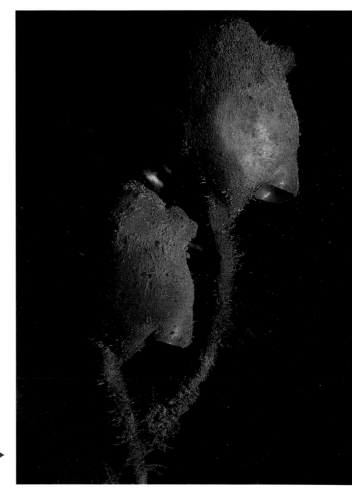

The phylum Chordata contains more than just the subphylum of vertebrates, and though it seems unlikely, we vertebrate humans are grouped in the same phylum as a collection of sea animals known as sea squirts. Creatures of the subphylum Urochordata include sea squirts, tunicates, and salps, which have been placed in the phylum Chordata because they have a stiff cartilaginous dorsal chord, called a notochord, at some time in their lives. This notochord is similar to a backbone and provides rigidity to the body. Any resemblance to other chordates ends there.

Sea squirts and tunicates are part of the class Ascidiacea. Sea squirts are roundish animals that live attached to a substrate, either directly or on a stalk of some type. The mature animal has two openings: an incurrent siphon and an excurrent siphon. The incurrent siphon opens into a pharynx, or branchial sieve, with slotted walls. By means of cilia, the animal moves water through the incurrent siphon, where the pharynx filters out tiny bits of organic matter for food by trapping it in a mucous coating. The water is then expelled through the excurrent siphon. The pharynx also extracts oxygen from the water.

Oddly, adult sea squirts do not have a notochord. It is found only in the larval "tadpole" stage, when they are planktonic and swim around. They look like tiny tadpoles, hence the name. As they mature, sea squirts lose their notochord and settle to the sea bottom.

The reproduction cycle in urochordates can be either sexual or asexual. Many species of these creatures form colonies by asexual budding. However, most species are also hermaphroditic and can reproduce sexually through a synchronized release of gametes from many individuals, with fertilization occurring in the water. A few species actually produce eggs that are incubated within the body of the sea squirt and released through the excurrent siphon after hatching into tadpoles.

The salps (class Thaliacea) are similar to sea squirts and tunicates but are planktonic throughout their lives. They function like tunicates and have two siphons, but the animal is shaped more like a

◄ *All tunicates, like these Sea Peaches, feed by filtering plankton from the water. Using cilia as a water pump, a Sea Peach draws water in through the larger opening and expels it through the smaller one.*

*The Stalked tunicate (**Boltenia ovifera**) grows on a stem.* ►
Unlike the Sea Peach, it cannot pump its own water.
When the current flows past, it bends the tunicate's stem
over, thereby forcing water in through its bottom siphon.

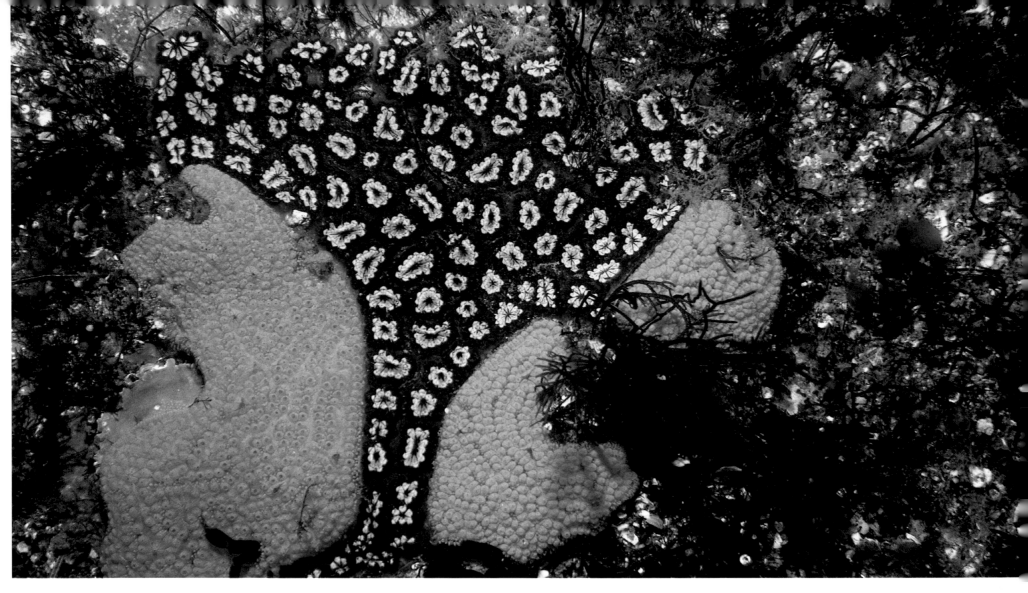

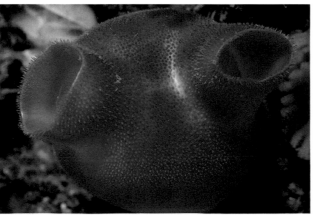

▲ *A colony of Golden Star tunicates (**Botryllus schlosseri**) growing within a colony of Orange Sheath tunicates (**Genus Botrylloides**). The Golden Star Tunicate is a favorite creature for use in the study of genetics, as they reproduce rapidly by asexual division. Orange Sheath tunicates are a Pacific species which were accidentally released into the North Atlantic in 1973.*

◄ *It may be hard to believe, but this animal, known as a sea squirt, is actually grouped into the same taxonomic phylum as humans. It is a chordate because during its larval stage it has a stiff dorsal chord, called a notochord, which functions somewhat like a backbone. This is the common Sea Peach (**Halocynthia pyriformis**).*

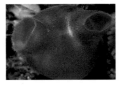

UROCHORDATES

barrel. One siphon is located on each end. Water moving through the siphons provides propulsion, as well as food and oxygen, to the animal.

The lancelet is another strange chordate in the subphylum Cephalochordata. This animal is about as closely related to vertebrates as any on Earth, without actually being a vertebrate. The small number of lancelet species have a notochord like the urochordates but actively swim. Lancelets prefer to dig into sand, where they feed and have been observed "swimming" through a sandy bottom as fast as a minnow swimming through water.

TUNICATES AND SEA SQUIRTS OF THE NORTH ATLANTIC

The most common sea squirt in the North Atlantic is the Sea Peach (*Halocynthia pyriformis*), which resembles its namesake in color and shape. This large ascidian really does look like a peach, especially when contracted. Sea peaches grow to 5 inches (12$\frac{1}{2}$ cm) in height and are found from Arctic to Massachusetts, although the larger specimens grow in the more northern reaches of their range. They have been found from shallow water to depths of over 600 feet (183 m). Like most other sea squirts, they have a larger incurrent siphon than excurrent siphon. Both siphons are located on the top of the animal. When disturbed, the Sea Peach contracts its siphons shut and shrinks to an almost perfectly spherical shape. They are extremely sensitive to motion in the water and will shrink up at the slightest disturbance. This makes them difficult to

photograph. Yellow and even completely white sea peaches are sometimes found.

The Stalked Tunicate (*Boltenia ovifera*) is similar to the sea peach, but with a few significant differences. It grows on a long stalk, up to 12 inches (30 cm) long, instead of attaching directly to the substrate. Other creatures may take advantage of the available real estate on this tunicate's stalk. Bryozoans, hydroids, algae, and barnacles will all grow well there. The incurrent siphon is located on the bottom of the animal, while the excurrent siphon points up. This inverted arrangement makes sense because the Stalked Tunicate is a passive feeder and cannot pump water through its siphons with cilia. It relies on ocean currents to bend the stalk and force water through the incurrent siphon, which then faces into the current. This is a fascinating example of evolutionary adaptation to an environment. Naturally, these tunicates are found only in areas of strong currents, like bays, where the tidal fronts create large water movements.

The main body of the animal can reach 3 inches (7 cm) in height, although specimens of this size are encountered only in water with swift currents and plenty of nutrients. They grow from the Arctic to Cape Cod in shallow water to depths greater than 1,600 feet (488 m). Eskimos eat this tunicate when it is blown ashore by storms.

Among the most well-studied tunicates in the world is the Sea Vase (*Ciona intestinalis*), which resembles a vase, both in shape and color. It is a translucent white with yellow coloring around the siphons. Growing to 6 inches (15 cm) in height, this tunicate is widely distributed from the Arctic to Rhode Island in the Atlantic and along the entire west coast of the United States. They are found to a depth of about 1,600 feet.

COLONIAL TUNICATES

The Golden Star Tunicate (*Botryllus schlosseri*) is a colonial, or compound, tunicate, which lives in a colony of hundreds or thousands of individuals, just like a hard coral colony. These animals generally reproduce asexually, effectively cloning themselves by the hundreds. Rapid asexual reproduction makes this tunicate a favorite subject for genetic research.

Individuals in a colony reach a height of only about 1/8 inch (3 mm), but the colonies may measure several inches across. They are found from the Bay of Fundy to North Carolina.

The Orange Sheath (*Botrylloides spp.*) is also a colonial tunicate. Although one of the most conspicuous tunicates in the southern Gulf of Maine, it was introduced here accidentally in 1973. A biologist brought some samples from the Pacific to the Woods Hole Oceanographic Institution for study. At the end of the summer, he placed samples on slides into Eel Pond as a holding tank. Yet these tunicates were so hardy that they multiplied and "escaped." In only twenty-three years they have populated practically the entire New England coast. A colony is usually only a few inches (7 cm) across but contains thousands of individual tunicates only 1/64 of an inch (<.5 mm) across. The animals are so small that the colonies look like sponges.

NEKTONIC INVERTEBRATES

Introduction

Benthic creatures live at the bottom of the sea. Nektonic creatures (called nekton) swim through the ocean. Usually nekton stay off the bottom and move in the water column. Fish, squid, and whales are all nekton.

Nektobenthic creatures leave the bottom occasionally but still spend most of their time on the ocean floor. Skates and flounders are nektobenthic creatures.

Within the broad group of animals we call nekton are two loosely defined subgroups. Animals of the open ocean: tunas, dolphins, whales, and others, which stay far out at sea most of the time, are called pelagic. Most nekton, however, are part of a second subgroup of coastal, or inshore, species. These creatures stay close to shore because they rely on the bottom either for food, shelter and protection, or mating.

Of the approximately 200,000 species of creatures in the ocean, only about 2 percent (about 4,000 species) are nektonic.

Most nekton are creatures with skeletons, such as fishes and mammals, but some invertebrates live as nektonic swimmers rather than benthic bottom-dwellers.

In the North Atlantic, squid exemplify the nektonic invertebrates. Octopods, which are found only in deep water in the North Atlantic, and squid are members of the class Cephalopoda, meaning 'head-footed.' Their "feet" (usually called arms or mistakenly called tentacles) grow from the part of the body containing the eyes (the "head"), while the rest of the body is out in front of the head. The body, therefore, does not connect directly to the arms.

Grouped within the phylum Mollusca, cephalopods are closely related to snails and clams. Though different in physical appearance, they share a similar internal construction. The most obvious difference between cephalopods and other mollusks is their apparent lack of a shell. Octopods do not have shells at all, and squid have a small chitinized internal shell. Nautiluses are the only cephalopods with an external shell, though they are found only in the South Pacific and Indian Oceans.

Cephalopods have the most highly developed nervous system of all mollusks, as well as the best-developed eye. The cephalopod's eye is a notable example of convergent evolution. It evolved from a completely different direction than the mammalian eye, yet it functions in almost the same way, giving cephalopods extremely good eyesight. With eyesight well suited for finding prey, cephalopods grasp their food firmly in their arms and eat using a mouth located between their arms. The mouth with a beak, similar to a parrot's, is used to bite into prey.

The octopus uses a salivary gland to secrete venom, which subdues its prey. The venom of the Blue-Ringed Octopus of the South Pacific is so powerful that its bite is almost always lethal to humans. Fortunately, it takes a lot of provocation before an octopus, any octopus, will bite a person.

While the octopus has eight sucker-equipped arms, the squid has ten. Eight of the squid arms are of the same length. The other two are extra long and grab its prey; these two long arms are the tentacles.

Cephalopods do share many characteristic molluscan traits, such as a mantle, a mantle cavity, a radula in the mouth, and a U-shaped digestive tract (useful for a creature in a conical shell).

SQUID

Squid move using two modes of propulsion. The fastest way is to squirt water out of the mantle cavity through a nozzle called a siphon. Because squid have a streamlined shape, they can move quickly with this jet-propulsion technique. A pair of fins on the forward portion of their bodies acts as stabilizers, and because their siphons can be pointed in different directions, squid can dart backward as well as forward. Squid can also hover or swim slowly, using their fins alone.

Squid, like octopods, can change color almost instantly. The skin of the squid is covered with small "bags" (called chromatophores), which are different colors. When certain muscles are contracted, different chromatophores can be expanded, producing colored dots on the skin approximately 1/16 inch across (1 mm). When those

muscles are relaxed, the chromatophores become nearly invisible. When approached, a cephalopod will change colors in an attempt to camouflage itself. Sometimes, colors can indicate a mood or temperament. The Atlantic Long-Fin Squid (*Loligo pealei*) turns deep red when bothered by divers but changes back to a blue or golden color when it does not feel threatened.

Octopods and squid share the ability to squirt a dye from their mantles. The dye is an extremely dark liquid, which is produced by a special gland. India ink was originally made by collecting the contents of this gland in cuttlefish (a type of squid) in the Indian Ocean. There has been much speculation about exactly what a cloud of dye in the water does for the squid (or octopus), but the cloud probably serves as a decoy for the squid while it escapes a predator, or it may confuse the predator's scent. Many squid live in deep water, where there is little or no light. These squid often produce bioluminescent ink, which glows dimly. This could certainly be a visual distraction to a predator in the depths.

SQUID OF THE NORTH ATLANTIC

The Atlantic Long-Fin Squid (*Loligo pealei*) is common in the North Atlantic, ranging from the Bay of Fundy to the West Indies. In mid-to-late summer, these squid go to shallow water to spawn and are easily seen by divers. They lay eggs in cigar-shaped masses attached to rocks and other stationary objects. They are swift swimmers but are also curious and will investigate a diver before jetting away,

occasionally in a cloud of ink. These squid are not a favorite for human consumption but are used sometimes as bait. Although tiny compared with Giant Squid, Long-Fin Squid are the largest of the squid species found in North American waters, growing to 17 inches (43 cm) long. They eat small fish, which they catch, hold in their arms, and then bite with their beak.

The Short-Fin Squid (*Illex illecebrosus*) is found in the North Atlantic from Greenland to North Carolina. These squid have a long and skinny shape but reach only about 1 foot (30 cm) in length. They observe breeding habits similar to those of the Long-Fin Squid, coming into shallow water during the spawning season, but otherwise they stay out at sea. One interesting observation about these squid is their apparent ability to jump completely out of the water with a powerful jet of water from their mantles. This may be a behavior similar to that of the so-called Flying Fish of the tropics: a method of escape from predation.

DIVING WITH SQUID

My first experience with squid was unexpected. On a night dive in a shallow Massachusetts cove in early July, I saw something move quickly just at the outer limit of my flashlight's beam. I swam in the direction of the motion and confronted a school of squid hovering in the water. There were more than a hundred light-blue squid looking at me, holding their ground. I was awed and did not dare breathe, since a plume of noisy bubbles might frighten the school away. Finally I chanced a breath,

NEKTONIC INVERTEBRATES

and as my exhausted air bubbled into the water, they flashed with color: reds, yellows, browns. I was diving with a 1:1 extension tube on my camera, and there was no hope of photographing so large a school of squid. Since I had nothing to lose, I decided to see how close I could get. I kicked my fins and moved a little closer. The school turned and began a slow swim, using only their fins. They swam gracefully and slowly enough for me to keep up. I followed along for quite a while, when suddenly, with a squirt of water, first one, then another, then the rest of the squid simply jetted out of my range of vision. I was lost and had to surface in order to figure out whether I was still in the cove. As it turns out, my swim with the squid had taken me all around the cove but not out of it. We had never gotten deeper than about 30 feet (9 m).

On another occasion that year I chanced upon a pair of mating squid. On that dive, I was shooting motion picture film. In my excitement to capture the squid lovemaking, I got a little too close. Suddenly, the pair separated, turned a dark shade of angry red and began ramming me. I was under attack! After a few seconds, the two squid jetted off to privacy elsewhere. What does a filmmaker know about modesty?

In the summer of 1992 we had a very long, cold spring and a cool summer. For material for an educational film, I dove repeatedly in the usually good spots for squid but could not find any. Even into August, there were no squid. Apparently the water was too cold (it never reached its usual low

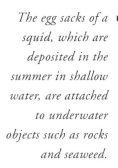

The egg sacks of a squid, which are deposited in the summer in shallow water, are attached to underwater objects such as rocks and seaweed. ▶

60s at the surface) for the squid to mate. With another diver, I headed down to Jamestown, Rhode Island, in search of warmer water. Being on the south side of Cape Cod, Rhode Island is sheltered from the cold water of the Labrador Current. Rhode Island also frequently receives very warm water from Gulf Stream eddies (see Introduction). The differ-

ence in water temperature between northern Massachusetts and Rhode Island is striking and can exceed 10°F at the surface. After a forty-minute dive, we stumbled into a small school of squid and got the footage we needed. The two minutes of usable footage we shot (in 5- to 10-foot visibility) proved to be the only squid that we saw all summer.

◀ *A pair of Atlantic Long Fin squid (**Loligo pealei**) prepare to mate in the waters off of Rockport, Massachusetts.*

◄ *Yellow Sea Ravens are rarely seen. It is likely that their coloring is "aposematic," meaning that it says to other creatures "I don't taste good." These fishes can inflate with water in order to appear larger than they are.*

FISHES

Humans and fishes are remarkably alike: we both have bilateral symmetry, a brain, a spinal chord, a heart, and other similar body organs. We both are members of the phylum Chordata and the subphylum Vertebrata. The key difference, however, is that humans adapted to life on land, while fishes adapted to living and moving through water. Most fishes are cold-blooded vertebrates with fins and breathe using gills.

Fish body design is optimized for a particular habitat within the ocean. Pelagic and schooling fishes are generally streamlined and muscular, able to swim quickly to escape capture or to pursue prey. Other fishes are flat so that they can lie on the bottom and patiently wait for prey to approach. Some fishes are camouflaged to resemble their habitat. The camouflage can be simple or extremely complex, depending on the species. Fishes vary in appearance and build, giving each type of fish a different place in the food chain. Each fish is unique and has a different method for survival.

 ## STAYING OFF THE SEA FLOOR

While nektobenthic fish are suited for life on the sea floor, pelagic fish swim continuously in the water column. The problem is that fish are more dense than water and tend to sink, so staying afloat is just as important to fish as it is to plankton.

Most fishes use a swim bladder to control buoyancy. Adding gas to the swim bladder allows increases buoyancy; taking it out again reduces buoyancy. It might seem that once the proper amount of gas is in the bladder, no further adjustment would be necessary. In fact, there is a complication. As a fish swims deeper, the water pressure around it compresses the gas within the swim bladder, causing it to decrease in volume and provide less lift. Either the fish adds more air, or it sinks. As a fish rises to the surface, it must remove air from the bladder because the decreasing water pressure allows the swim bladder to expand and could literally cause the fish to explode.

Analysis of the gas in fish swim bladders shows that most shallow-water fishes use a mixture of gas resembling air (roughly 20 percent oxygen and 80 percent nitrogen). This may seem logical,

and it is, but not necessarily for the reason many people believe. Although a few species can gulp air at the surface, most fishes must fill and empty their swim bladders through their gills. Fish use a gas gland in the swim bladder to exchange blood gasses. Adding gas to the swim bladder means first taking gas from the water through the gills, getting it into the bloodstream, and then passing it into the gas gland, which finally fills the swim bladder. This exchange takes time and limits the rate at which a fish can change its depth. For that reason, fishes that are caught in the deep sea and hauled quickly to the surface can appear deformed by the rapid and uncontrolled expansion of the swim bladder

Many extremely deep water fish evolved a different kind of swim bladder. At depths greater than 16,000 feet (5,000 m), the density of the gas in a swim bladder is nearly the same as the density of fat (but still less than muscle tissue). Some deep-water fish have evolved a swim bladder filled with fat instead of gas. The advantage of fat is that it is almost nearly incompressible, and therefore buoyancy is not altered by by a change in depth. The disadvantage is that the fish cannot change the amount of buoyancy provided by the swim bladder.

◀ *A Sand Tiger shark (**Odontaspis taurus**) uses a school of small fishes as cover
to evade detection by larger fishes which the shark hopes to catch. The smaller
fishes enjoy protection from their predators while swimming with the shark.*

BODY TEMPERATURE

Although most fishes are cold-blooded (that is, their body temperature is about the same as the water around them), there are some warm-blooded exceptions. One notable example is the Bluefin Tuna (*Thunnus thynnus*), which maintains a body temperature of between 5° and 20°F higher than the surrounding water. It is thought that the Bluefin's higher body temperature helps to increase its metabolic processes, allowing the muscles to produce the increased power it needs for sustained high-speed swimming.

CLASSES OF FISH

There are only three classes of living fishes: the class Agnatha, jawless fish (the lampreys and hagfishes); the class Chondrichthyes, cartilaginous fish (sharks and skates); and the class Osteichthyes, the bony fish.

CLASS AGNATHA

These fish have no jaws. Instead, their mouths are designed to suck body fluids from other creatures. Like sharks and skates, the agnathan skeleton is formed from cartilage, not bone, and the skin has no scales. Lampreys are found in both freshwater and marine habitats but spawn only in freshwater.

The only common member of the class of jawless fish in the North Atlantic is the Sea Lamprey (*Petromyzon marinus*). As a larva, it is a filter feeder and lives in the mud for up to five years. As an adult, this parasitic lamprey attaches itself to other fish. The Sea Lamprey is an anadromous fish, spawning in freshwater and then migrating to salt water. In mature form, they reach up to 33 inches (84 cm) and are found from the Gulf of St. Lawrence to Florida.

 CLASS CHONDRICHTHYES

The cartilaginous fish are organized into three groups. All have one trait in common: the absence of a bony skeleton and the presence of a cartilaginous one. The least well known group is the Ratfish (family Chimaeridae), whose only known member in North America is found on the Pacific Coast.

The other two groups of cartilaginous fishes are the sharks and the batoids, which include skates, rays, and sawfishes. Batoids have ventral gill slits and pectoral fins that are fused to the side of the head to form a disk-shaped body. Sharks have gill slits that are at least partially visible from the side.

None of these creatures has swim bladders. The sharks and pelagic rays must keep moving in order to stay up in the water column, like an airplane in the sky. The skates just stay on the bottom.

SHARKS

Perhaps no other ocean creature strikes fear into the hearts of swimmers as the shark does. On the whole, however, sharks suffer from an undeservedly bad reputation. Statistically, few humans have been attacked by sharks, certainly fewer than have been bitten by dogs. Yet we still label sharks man-eaters and killing machines. The truth is that they are killing machines, but people are not on their list of preferred foods. Putting their faith in statistics, most scuba divers consider it a thrill and a privilege to see a shark underwater. Most find that it is hard to get close to a shark because the sharks are so timid.

Sharks, like most predators, are relatively lazy and prefer to stalk food that offers little resistance when attacked. Sharks tend to prey on animals that are either sick or appear disabled. One explanation for shark attacks on people is that most swimmers are not very graceful (compared with healty fish) and appear "sick" to the shark.

Attacks by sharks on surfers in California and Australia may result because a person paddling a surfboard looks to a shark like a seal swimming at the surface. Seals are a favorite food of sharks, especially Great Whites. Seals can outswim sharks, unless one is sick or caught unaware. In many documented cases where sharks or orcas ('killer whales') have attacked surfers, the predator has spit out the surfer once it realized that the human did not taste like a seal.

▾ *The Spiny dogfish (**Squalus acanthias**), a small shark found in coastal waters all over the North Atlantic. It is not a threat to people.*

The role of sharks in the ocean's ecosystem – like that of the largest carnivores on land – is a significant one. By feeding on the sick and weak, sharks help to maintain a healthy population of smaller animals and keep the gene pools free of disease. Lions, tigers, and wolves perform the same roles in their respective ecosystems.

As a predator, the shark is nearly perfect. Inside a shark's mouth are rows of razor sharp teeth. As teeth are worn out or broken, a new tooth from the next row grows in to replace it. The shark's sense of smell is advanced. Experiments prove that some sharks can detect a drop of blood in 1 million gallons of seawater. Using this powerful sense, a shark can home in on an injured fish from a substantial distance.

Recent studies have also shown the shark's eyesight to be acute. Behind the retina of a shark's eye is a mirror that reflects light back through the photoreceptors a second time, dramatically improving the shark's low-light vision. When it cannot see, it picks up vibrations in the water by sensors around its body. These sensors help steer the shark in the direction of prey, even in pitch darkness. In addition, sharks are acutely sensitive to electrical currents, and the impulses emitted by the muscles of a fleeing or struggling fish provide a homing beacon for a shark in pursuit.

Unfortunately, their bad reputation puts sharks in danger of extinction. Hunted ruthlessly by humans for sport, many species are now endangered, including the Great White. Certain Asian recipes call for shark fins, which are considered a

 # FISHES

delicacy. In Japan, sharks are hunted just for their fins and are thrown back alive to die without them. (This cruel process, known as finning, was recently outlawed in the United States.) If sharks are to survive in today's ocean, we need to understand their importance and protect their place in the ecosystem.

Brian Skerry, a friend and fellow photographer who provided many of the shark photographs in this book, has been leading shark-diving expeditions for years. By his own admission, he is a "shark fanatic." To take photographs of Blue Sharks, Brian charters a boat out of Rhode Island, casting off before dawn. His shark-filming equipment includes an antishark cage, several buckets of chum (chopped fish), and a whole lot of patience. The day begins with a 3- to 4-hour cruise to reach areas 40-50 miles offshore, where the sharks congregate. Then, he chums for several hours, in order to leave the smell of blood in the water. This messy business attracts sharks and is usually the only way they can be persuaded to stay long enough for a photo session. When the sharks finally arrive, the divers get into the water and begin filming. Brian has a rule of thumb about the antishark cage: if there are fewer than three sharks, it is not necessary. If there are more than three, it becomes hard to keep visual track of each animal, and so the divers use the cage for safety.

Sharks are not there to eat divers. They are interested in the fish smell and spend their time swimming through the chum, eating tasty fish morsels. Brian has never been attacked by a shark.

REPRODUCTION IN SHARKS

Sharks can reproduce in one of three ways. All sharks copulate and then fertilize internally. What happens after that depends on the species. Oviparous species lay large, tough eggs on the bottom, which incubate for up to a year. The egg contains a large yolk sac that provides nourishment for the developing embryo.

Viviparous sharks give birth to young much as mammals do. They have a uterus in which one to as many as one hundred embryos develop, each nourished by an umbilical cord from the mother. The embryos of several species of sharks are known to engage in prenatal intrauterine cannibalism, in which the strongest or first born of the embryos will eat other embryos or unhatched eggs to nourish themselves.

Ovoviviparous sharks combine the two approaches. The mother produces eggs, but the eggs are held internally until they hatch. During birth, ovoviviparous sharks have the appearance of viviparous sharks, but there is no uterus in the mother.

SHARKS OF THE NORTH ATLANTIC

Several species of sharks live in the North Atlantic, but most of them are not considered dangerous. Some are plankton eaters with no teeth at all, while others are small and skittish, with no desire to make a meal of a person.

THE CARPET SHARKS

The Nurse Shark (*Ginglymostoma cirratum*), found from Rhode Island to the Gulf of Mexico and throughout the Caribbean, is a member of a family of sharks known as Carpet Sharks (family Orectolobidae). Carpet Sharks have several distinguishing characteristics, including two dorsal fins of equal size that lack spines and well-developed barbels ("whiskers" made of flesh). The Nurse Shark is the only Carpet Shark species found in the North Atlantic. It grows to 14 feet (4 m) long and is most abundant in shallow water off of the southern coasts of Florida. It is harmless to people and prefers a diet of sea urchins, small fish, and crustaceans. The Nurse Shark is sometimes referred to as a suction shark because it inhales its prey like a vacuum cleaner.

THE WHALE SHARK

The largest fish in the sea is actually a shark. The enormous (and completely harmless) Whale Shark (*Rhincodon typus*) can grow to 60 feet (18 m). This shark is so unusual that it is grouped in its own family (Rhincodontidae). What makes this animal so unusual, besides its size, is its preferred diet of plankton and small fishes, which are strained from the water using a branchial sieve. Although primarily a warm-water shark found in the tropics, it can occasionally be seen as far north as New York.

*A Blue shark (**Prionace** ▶ **glauca**) in the open ocean, several miles from the Rhode Island coast.*

THE SAND TIGER

The family Odontaspididae contains the Sand Tiger sharks, of which there is one species in the North Atlantic, simply called the Sand Tiger (*Odontaspis taurus*). This shark can reach 10 feet (3 m) in length and is found from the Gulf of Maine to Brazil. They are most abundant north of Cape Hatteras, North Carolina.

Many people claim that the Sand Tiger is dangerous, but that assumption is based more on appearance than experience. Its mouth is filled with sharp, exposed teeth, which gives it the look of an animal always ready to strike. This shark is actually tolerant of divers and is one of the easiest sharks to maintain in captivity because of its mellow disposition. It is, therefore, the ideal shark to photograph: it looks mean on film, but in real life it is not mean. Incidentally, the nasty-looking teeth of the Sand Tiger are long and thin and point into the mouth slightly. They are not designed for cutting off huge chunks of flesh from a large prey animal, but instead for holding on to smaller prey, like slippery fish.

THE MACKERAL SHARKS

The Mackeral Sharks (family Lamnidae) are best known for the Great White, probably the most dangerous shark in the seas. The Basking Shark (*Cetorhinus maximus*) is also part of this family and is as harmless as the Whale Shark. The most obvious distinguishing characteristic of the family is simply the torpedo shape of the body.

The Great White (*Carcharodon carcharias*, also called the White Shark) grows to 21 feet (6.5 m) and is known for its role as the antagonist in the film *Jaws*. Great Whites are the most feared sharks in the world, although their extreme rarity leads marine biologists to consider the species on the verge of being endangered. They are found in oceans around the world in small numbers; in the Atlantic they range from Newfoundland to Brazil. Their favorite foods include large animals like seals, sea lions, and otters. They also eat fish and an occasional crab.

The Basking Shark is the second largest shark in the world, reaching lengths of 45 feet (14 m). This shark is 6 feet (2 m) long at birth and reaches 10 feet (3 m) by its first birthday. Like the Whale Shark, this animal strains plankton from the water for food, has only tiny teeth, and is completely harmless to people. Basking Sharks frequently sun themselves at the water's surface.

The cooler temperate waters and the variety of plankton it has to offer attracts Basking Sharks to the North Atlantic. This makes the North Atlantic and Pacific Oceans the regular feeding grounds of these giants. In the Atlantic, they are found from Newfoundland to Florida.

Unfortunately, like the great whales, Basking

FISHES

Sharks were aggressively hunted in the past for liver oil and are only now beginning to make a comeback. The animal has an enormous liver. A 30-foot (9 m) Basking Shark may weigh about 6,500 pounds (2,950 kg), and the liver alone of that fish will weigh approximately 1,800 pounds (816 kg). This large liver serves as a sort of swim bladder, since it is less dense than water. Fortunately, the liver oil was only sought for its use in oil lamps and is no longer in demand.

During the summer, anglers off the shores of Massachusetts and Rhode Island frequently spot Basking Sharks cruising the surface waters, straining plankton. Sometimes they can be seen from shore, causing a bit of a commotion until it is clear that the huge sharks are harmless (unless you are a copepod). Basking Sharks seem to migrate north from North Carolina, since they are seldom seen south of there, and finally reach Maine by August. After the summer migration, no one knows where they go in winter. One theory suggests that they "hibernate" on the bottom.

The Shortfin Mako (*Isurus oxyrinchus*) resembles the Great White but is smaller, reaching only 13 feet (4 m). It feeds on small fish and is found from Cape Cod to Argentina in the Atlantic. Shortfins are considered dangerous and aggressive but are relatively rare (though not nearly as rare as the Great White). They have the distinction of being the fastest sharks in the seas because of their warm-blooded matabolism. A fully-grown mako can swim at 60 m.p.h.

THE REQUIEM SHARKS

The ocean's largest family of sharks, Carcharhinidae, more commonly called Requiem Sharks, includes twenty-six known species that occur in North American waters alone. Most of these sharks are pelagic and are not found on the bottom or near shore.

The most familiar Requiem Shark in the North Atlantic is the Blue Shark (*Prionace glauca*), which grows to more than 12 feet (3 .75 m) long. This cold-water shark is most often found from Nova Scotia to the Gulf of Maine, although it does range as far south as the Chesapeake Bay. Blue Sharks can be aggressive around divers and chum but are not usually dangerous to humans. They prefer to feed on small schooling fishes. Although implicated in the attacks on shipwreck victims during World War II, there is no conclusive evidence to confirm this accusation.

Many sharks have demonstrated the capability to sense the Earth's magnetic field. This trait may help sharks to navigate long distances. Although little is known about shark migration patterns, Blue Sharks do make large annual migrations over thousands of miles and may rely on these magnetic fields.

In comparison to Blue Sharks, the Tiger Shark (*Galeocerdo cuvieri*) is a rather indiscriminate eater. Tiger sharks are also Requiem Sharks, but unlike the Blue Sharks, their stomachs have been known to contain fish, dolphins, dogs, garbage,

turtles, and other strange foods. In the Caribbean, they are considered dangerous to people, although there are few documented attacks of Tiger Sharks on people. These animals reach 18 feet (5¹/₂ m) in length and can weigh 1,800 pounds (816 kg). They are found from Cape Cod to Argentina.

THE DOGFISHES

The family Squalidae contains the Dogfishes, or Dogfish Sharks. The family name comes from a Latin word meaning rough or scaly, which describes the rough skin of these sharks. In the North Atlantic, the Dogfishes are represented by the relatively common Spiny Dogfish (*Squalus acanthias*). This small, skittish shark reaches only 4 feet (3.25 m) in length. Its name comes from the single spine on the leading edge of each of its two dorsal fins. These spines, used for defense, are mildly venomous.

Tagging studies show that the species is migratory, an assumption widely held by New England divers. The sharks pass through local dive spots at about the same time every summer and prompt a flurry of interest among divers anxious to see them. After a week or two, they are gone. No one seems to know where they go. They are not hunted or considered dangerous in the United States, but they are eaten in Europe. These sharks prefer water 30-1,000 feet (9-305 m) deep and are found in the Atlantic from Newfoundland to North Carolina.

◄ *The Little skate (**Raja erinacea**) is common in shallow coves all over the North Atlantic, from the Gulf of St. Lawrence to North Carolina.*

THE BATOIDS

The Batoids in the North Atlantic are represented mostly by skates and rays. Both animals are easy to identify because of their flattened winglike shape and mouths hidden on the animal's ventral or underside surface.

THE ELECTRIC RAYS

The family Torpedinidae contains the electric rays, which are similar to electric eels in that they can generate an electrical charge to stun prey and protect themselves. The animals have specialized muscles near their heads that produce the electrical charge.

The Atlantic's only species of electric ray, the Atlantic Torpedo Ray (*Torpedo nobiliana*), grows to 6 feet (2 m) in length and can produce about 220 volts. It is primarily a fish eater and is able to stun prey that it could not otherwise catch, since it does not swim particularly fast. The Torpedo Ray looks a bit like a pancake, as its body is round and flat. It is found from Nova Scotia to Florida.

 ## THE SKATES

The family Rajidae includes skates, which are similar to rays, but are almost always nektobenthic. Skates propel themselves in one of two ways. They can swim by flapping their "wings" like rays, or they can use a pair of pelvic fins beneath and toward the rear of their bodies. The animal "hops" along the bottom using these fins, while extending its wings to allow gliding. This is the most common way that the animal searches for food. In addition, males have a pair of clasper appendages on the pelvic fins that are used for mating. Skates do not have a stinger at the base of the tail like a stingray. They are completely harmless and feed on animals found on the bottom, like sand dollars, bivalves, and worms.

Among the seven known species of skates in the North Atlantic, the most common is probably the Little Skate (*Raja erinacea*). At one particular cove in Massachusetts, these skates are so numerous that on a night dive during the summer it is possible to count a hundred of them. The cove probably attracts so many skates because sand dollars (a favorite food) are abundant there.

When a Little skate senses the approach of a diver, it buries itself in the sand with a few quick flaps of its pectoral "wings." The skate relies on its color and the partial sand cover to serve as camouflage. If, after waiting for a time, the diver does not leave, the skate swims casually away. Although it could outswim a diver with a burst of speed, it will instead take a leisurely cruise along the bottom with the diver in pursuit. If the diver gets too close, the skate employs its next defense. It swims around in a 10-foot (3-m) diameter circle and finally disappears in a cloud of silt, leaving the diver, as it were, in the dust.

The Leopard Skate (*Raja garmani*) is another common North Atlantic skate. It resembles the Little Skate in general shape and size, except that it has a striking leopardlike pattern on its dorsal side.

*Lying flat on the ocean floor ▶ allows flatfishes like the Windowpane Flounder (**Scophthalmus aquosus**) to blend in quite effectively. This is a lefteye flounder.*

THE BONY FISHES (CLASS OSTEICHTHYES)

On the evolutionary calendar, bony fishes are children of the Jurassic period, only some 150 million years old. Sharks are graybeards by comparison and claim ancestry stretching back 300 million years. Bony fishes, named because all or part of their skeleton is bone, have proved immensely successful and are now the most numerous species in the world's oceans. Unlike sharks, they have a single pair of gill openings and usually a swim bladder.

Most bony fishes do not range as widely as sharks, so many species have evolved to survive in a particular habitat or on a particular diet. Sometimes, this specialization has produced many closely related, yet different, species. Such high levels of specialization made some species of fish intolerant of even small changes in their environment.

Because of the incredible number of species of fish in the North Atlantic, it would be difficult to discuss them all, but I include here a group that seems particularly appropriate.

◀ *The fleshy tabs hanging off of the Sea Raven (**Hemitripterus americanus**) help it to blend into its habitat of seaweed on the bottom.*

 FLOUNDERS

Flounders are flatfish, which have evolved to lie flat on the ocean floor. In addition to being perfectly flat, the flounder also has both eyes on the same side of its head. This preposterous twist of evolution allows the flounder to keep a close watch for predators while on the bottom. Unlike skulpins and other bottom-dwelling fish that lie on their bellies, flounders lie on their sides, which is why both eyes are on the same side.

Interestingly, flounders do not begin life looking at all the way they do as adults. A flounder grows from an egg into a planktonic larval stage, where it floats around in the water column. As it matures, it begins to look like a tiny fish but has its eyes on opposite sides of its head. When the fish grows a little older, one eye migrates to the other side of the head, and a startling metamorphosis occurs in the animal's morphology as it changes from a normal-looking fish to one that is flat. Interestingly, the mouth stays in its original position and is vertical when the animal lies on the bottom.

All flounders are part of the Order Pleuronectiformes (meaning 'side swimmers'), but there are four families of these fish within the order: the Lefteye Flounders, the Righteye Flounders, the Soles, and the Tonguefishes.

Lefteye flounders (family Bothidae) are so named because when the fish matures, the right eye migrates to the left side, where both eyes remain in the adult. In addition, all the coloration is on the left side, which, obviously, is the side that faces up

FISHES

when the animal lies on the bottom. The right side, therefore, becomes the underside and takes on a light color.

The most common Lefteye flounder in the North Atlantic is the Windowpane Flounder (*Scophthalmus quosus*), which can grow to 18 inches (46 cm) but is usually smaller. These fish prefer sandy bottoms, where they blend in well. They are found from the Gulf of St. Lawrence to Florida and are common in New England but are not usually eaten or taken as game fish.

Righteye flounders (family Pleuronectidae) have eyes and coloration on their right sides. It is easy to distinguish a Righteye from a Lefteye flounder by looking at the orientation of the mouth, gill slit, and pectoral fin with respect to the eyes. A Lefteye Flounder viewed from the "top" clearly looks like the left side of a fish with a second eye placed on its head, while the reverse is true for the Righteye flounders.

The largest flounder in the North Atlantic, and probably the second largest in the world, is the Atlantic Halibut (*Hippoglossus hippoglossus*). This gigantic Righteye flounder has been recorded at more than 700 pounds (317 kg) and 8 feet (2.5 m) in length, although specimens between 100 and 300 pounds (45 and 136 kg) and 5-6 feet (1.5-2 m) in length are more common. In the North Atlantic,

only Ocean sunfish, tunas, swordfish, and a few sharks grow as large.

Atlantic Halibuts prefer to eat fish and rarely consume invertebrates, except when fish are scarce. They range from Greenland to New York, preferring cold, but not arctic, water. Fishers report that the largest specimens are found in water between 36° and 38°F. They seem most numerous at depths of around 300 feet (191 m).

Although it is not certain how long these large animals live or how fast they grow, it is likely that they spend about a year drifting in the current as planktonic fish fry before they metamorphose into adult forms. Evidence suggests that females reach sexual maturity at 9-10 years. Atlantic Halibut may live 60-70 years if they can manage to avoid the nets of trawling fishers.

Although not as large as the Atlantic Halibut, the Winter Flounder (*Pseudopleuronectes americanus*) can grow to just under 2 feet (0.6 m). This fish gets its common name because of the water temperature it prefers. Although the fish is found from Labrador to Chesapeake Bay, it is most abundant in the Gulf of Maine. South of the Gulf of Maine, the fish migrates to deep, cold water during the summer and returns to shallow coastal water only in the winter, when it finds the water temperature more comfortable.

The Soles (family Soleidae) are Righteye flounders that generally prefer shallow coastal waters. The family name can be confusing because there are many members of the Righteye (Pleuronectidae) family that are called soles yet are not included in the sole family.

The Naked Sole (*Gymnachirus melas*) is the most common North Atlantic sole but reaches only 6 inches (15 cm) in length. It is found from Massachusetts to Florida.

▲ *The Shorthorn Skulpin (**Myoxocephalus scorpius**) is very common throughout the North Atlantic, preferring extremely cold water.*

◄ *The Winter Flounder (**Pseudopleuronectes americanus**) has its eyes located on the right side of its body, and is there fore called a righteye flounder.*

 # FISHES

 ## THE SKULPINS

Members of the skulpin family (Cottidae) range throughout the world's oceans. They are closely related to the scorpionfishes (family Scorpaenidae), which are familiar to many divers in the tropics who learn to avoid their extremely venomous spines. Although skulpins resemble scorpionfishes, their long, sharp spines are generally not venomous. The cold polar oceans are populated with many different species of these bottom-dwelling fishes.

One of the most unusual looking skulpins in the North Atlantic is the Sea Raven (*Hemitripterus americanus*). This strange-looking fish has rough skin and many fleshy tabs sticking out in all directions. Together with a variable mottled coloring, it blends in well with the surrounding sea floor. Sea ravens have many small teeth that are capable of delivering a bite, but these fish actually tolerate divers who photograph and, in many cases, handle them. When disturbed, they can inflate themselves with water. This reaction to harassment makes them appear larger but renders them relatively immobile.

Although most Sea Ravens vary between brown and red in color, the lucky observer may sometimes find a yellow example. They can reach a little more than 2 feet (0.66 m) in length, weigh close to 10 pounds (4.5 kg), and range from the Gulf of

◄ *The smallest skulpin in the North Atlantic is the Little skulpin (**Myoxocephalus aeneus**). It is related to the extremely venomous Scorpionfish found in the tropics, but is not venomous itself.*

Putting its faith in ▶ camoflage, the Sea Raven holds very still even as the camera approaches closely.

St. Lawrence to Chesapeake Bay in relatively shallow water (generally fewer than 200 feet, or 61 m, deep).

Less exotic, but equally common in the North Atlantic, is the Shorthorn Skulpin (*Myoxocephalus scorpius*) and Longhorn Skulpin (*Myoxocephalus octodecimspinosus*). These skulpins are similar in shape and size but are easy to tell apart if one knows where to look.

Both these fishes are found year round in shallow (usually fewer than 100 feet, or 30 m)

North Atlantic waters. They live in one area for long periods, not ranging far from home, and feed on crustaceans, mollusks, and worms found in the bottom. Unaffected by changes in the water temperature, they seem to prefer the cold, winter water. They hug the substrate, gliding close to the bottom with their large pectoral fins stretched out like wings. Both these skulpins grunt occasionally at the surface when caught by anglers, but divers frequently hear them grunting in the water as well.

◂ *The toothy smile belongs to a Goosefish (**Lophius americanus**). This depressed (i.e. flattened) fish suspends a fleshy lure over its mouth and waits for other fish to attempt taking the bait. With that mouth, the Goosefish can eat fishes as large as itself, as well as birds and turtles!*

THE GOOSEFISHES

Goosefishes (family Lophiidae) are flat, bottom-dwelling fish, sometimes called anglerfishes. The name comes from their ability to "angle" for food using a fleshy "lure" on a long spine (like a fishing pole) hung out over their mouths. With any luck, smaller fishes are attracted to the lure, called an illicium, and fail to see that it is attached to the goosefish, which is cleverly camouflaged to match the bottom. The North Atlantic Goosefish (*Lophius americanus*) is a formidable looking creature, with a gigantic mouth full of teeth permanently fixed into a toothy grin. This fish is flattened ventrally and blends in well with the bottom, where it waits for prey attracted to the lure, or to pass by unknowingly. It eats creatures as big as itself, including other fishes, birds, and turtles. The North Atlantic Goosefish reaches 4 feet (1.2 m) in length and is found from the Bay of Fundy to Florida from shallow water to depths of about 1,200 feet (366 m).

*An adult male Lumpfish (**Cyclopterus lumpus**) guarding ▸ his nest of eggs. The male Lumpfish will stand guard without eating for many weeks, driving away all intruders.*

 # FISHES

 ## THE LUMPFISHES

Lumpfishes (family Cyclopteridae) look unusual because of their ventral "suction cup," called a sucking disk, which serves as an anchor to hold on to rocks and other hard surfaces.

There is only one common lumpfish in the North Atlantic, cleverly called the Lumpfish (*Cyclopterus lumpus*). It is found from Newfoundland to Chesapeake Bay in shallow water and prefers to live near rocky bottoms. Lumpfishes can reach almost 2 feet long, although they are usually less than 16 inches (40 cm) long. They eat a wide variety of food, including small crustaceans (amphipods and isopods), worms, soft-bodied mollusks, jellies, and ctenophores.

The Lumpfish's reproduction cycle is unusual. The female can produce more than one hundred thousand eggs. She lays them during early spring in a spongy nest, generally under a rock overhang or some other protected place, and then swims away. The male then guards the eggs, never leaving the nest for any reason except to drive away intruders. He does not eat at all. Throughout the incubation period, the male gently fans water over the eggs to keep silt from settling on them, and to keep water moving through the nest. When the eggs finally hatch, the male lumpfish's job is over, and he goes in search of food. This story of courage and devotion ends happily because the male, though emaciated and exhausted, lives to breed another year. The planktonic larvae of the newly hatched eggs, however, must face many challenges to survive and reach adulthood. By the end of their first summer, the few individuals that have escaped predation are no more than an inch (2.5 cm) long. Lumpfish reach sexual maturity at three to four years, and a 12-inch (30 cm) adult fish is at least five years old.

Some aficionados find the taste of lumpfish eggs similar to caviar. In these days of overfishing and diminishing sturgeon and salmon stocks, many people now choose lumpfish eggs over caviar, a practice common in Europe and now becoming popular in the United States. At least one university is researching lumpfish aquaculture to learn how to farm these eggs.

While filming these interesting fishes, I encountered a male guarding his nest of eggs. He tolerated my camera for a moment, but as I drew closer for a macroshot of the nest, he tried to frighten me away by swimming toward me at full

▲ *Close-up of the Lumpfish eggs.*

speed and then veering off at the last minute. Although he never actually rammed me, I was certainly alarmed. I admired his courage in defending the nest from an intruder so much larger than himself. Eventually, he decided that I was not going to hurt the nest and resumed fanning his eggs, even though I stayed to film him.

Another diver reported being "butted" by a lumpfish when he inadvertently swam too close to its nest. He claimed that the lumpfish attacked him only when he wasn't looking!

◄ *Although this one inch long fish looks just like a juvenile Lumpfish, it is actually an adult Spiny lumpfish (**Eumicrotremus spinosus**). The Spiny lumpfish is rarely encountered in the Gulf of Maine, as they prefer the cold arctic waters far north of New England.*

FISHES

 **EEL-LIKE FISHES**

There are many families of eel-like fishes, which are some of the most interesting creatures in the North Atlantic. In general, eel-like fish have long, tapered bodies with a dorsal fin that runs the length of the body and a head looking like a fish. These are actually fish and not eels. True eels have no spines in their dorsal fins, while fish do.

Perhaps the most common of these fishes in the North Atlantic is the tiny Rock Eel (*Pholis gunnellus*). This small fish rarely reaches longer than 6-8 inches (15-20 cm), although a specimen 12 inches (30 cm) long has been recorded. The Rock Eel likes to hide under rocks in shallow water (a few feet, or 1 m, deep). They range from Newfoundland to Cape Cod and are an important food for cod and pollack.

The Oyster Toadfish (*Opsanus tau*) belongs to the family Batrachoididae and lives in the warmer areas of the North Atlantic. This fish has the ability to croak like a frog. It grows to about 15 inches (37.5 cm) and uses powerful jaws to crush mollusks and crustaceans. Oyster toadfish are found hiding in holes on the bottom from Cape Cod to Florida. Fleshy tabs all over these fish help them to blend in with their surroundings, sometimes making them nearly invisible.

Though small, the Feather Blenny (*Hypsoblennius hentzi*) is both handsome and interesting to watch. This 4-inch (10-cm) fish makes its home in a hole, like most of the eel-like fishes, and rarely leaves, except to hunt prey. Unlike larger fishes, however, the blenny does not search far for food. When prey is in range, the little blenny pops out, grabs its victim, and quickly retreats to the safety of its home. When a diver appears, the fish disappears into its hole, but patience on the diver's part usually pays off. The blenny frequently comes back a moment later and resumes the watch for prey, keeping a careful eye on the diver.

The Feather Blenny gets its common name from the pair of featherlike appendages (called cirri) on the top of its head. Their purpose is not known. Although this blenny is a dull brown color, males have a brilliant blue spot on the anterior part of their dorsal fin.

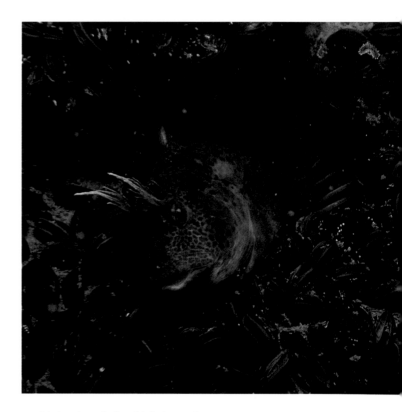

▲ *Living in a hole which it rarely leaves except to eat, the Feather blenny (**Hypsoblennius hentzi**) gets its name from the pair of feather-like appendages on its head.*

◄ *A Rock eel (**Pholis gunnellus**).*

FISHES

 **THE EELPOUTS**

The family Zoarcidae contains the Eelpouts, which are predominantly cold-water eel-like fishes. The most common North Atlantic Eelpout is the Ocean Pout (*Macrozoarces americanus*). This brownish-yellow fish grows to more than 31/2 feet long (1 m). It likes to hide in holes with only its head protruding to watch for intruders. It can coil its long body inside a small hole, which makes the fish appear smaller. When bothered by repeated flashes from a photographer's strobe unit, this fish will leave its retreat and swim off into deeper water. Watching one of these long-bodied fishes come out of a hole is like watching a long train emerge from a tunnel. The Ocean Pout is found from the Gulf of St. Lawrence to Delaware to about 600 feet (183 m) deep.

◄ *The Ocean pout (**Macrozoarces americanus**) is a long eel-like fish which thrives in the cold waters of the North Atlantic.*

 WOLFFISHES

The family Anarhichadidae includes the eel-like wolffishes, of which there are two species in North America: one in the Pacific and a different one in the Atlantic. In the North Atlantic, this ferocious-looking fish is simply called the Atlantic Wolffish (*Anarhichas lupus*). These fish are generally blue in color and are shaped like eels, with a long body and dorsal fin running the length of their backs. They have no pelvic fins. The head is enormous in comparison with the body, which is a sight at lengths of up to 4-5 feet (1-1.5 m). Its mouth is full of teeth: sharp canines at the front that resemble those of a snarling dog and molarlike teeth at the back.

The wolffish looks forbidding and when caught on a fishing line can be extremely dangerous. As it attempts to cut the line, it can bite off fingers with little effort. In fact, a wolffish can bite through steel fishing leader. In its natural element, however, the wolffish's teeth are used to consuming hard-shelled mollusks and echinoids. The wolffish is actually shy and withdrawn, despite textbook claims that it is fierce and dangerous to swimmers. Those reference books also say that the wolffish is a solitary animal, never found in groups. Both assertions are wrong but have given the wolffish an undeserved reputation almost as unsavory as that of the Great White Shark.

Oceanic Research Group filmed wolffish for an educational film. We spent hours with these fishes and came away knowing the personalities of many individual fish.

Many years ago on a dive out at Halfway Rock, off Marblehead, Massachusetts, I encountered my first wolffish in 90 feet (27.5 m) of water. I turned a corner and came face-to-face with a wolffish that was about 3 feet (1 m) long. I had never seen one before and did not know what it was. I did notice the large teeth, though, so I approached it carefully. We watched one another for a while, but nothing happened. It was a staring contest between human and fish, and the fish was winning. Unbeknown to me, I had stumbled on a rare find in the North Atlantic: a solitary wolffish out on the sand during the day. Since nobody was making a move, I gently prodded the fish with a finger. It backed up a little and kept watching. Finally, I got bored and swam away.

When I described the fish to the captain of the boat, he told me I had found the fiercest fish in the ocean and that I was lucky it did not bite my whole hand off when I touched it. Though the fish had seemed harmless, I nevertheless made a mental note not to fool with big, blue, pointy-fanged fish.

 **FISHES**

◄ *Atlantic Wolffishes (**Anarhichas lupus**) are frequently found in pairs. This is because once the Wolffish has found a mate, the pair remains together for life!*

Years later I read about a place in New Brunswick frequented by wolffish. I traveled there with Tom Krasuski, and we spent a week searching but found no wolffishes. Two months later, we undertook a second trip. With advice from local divers, we found the magic wolffish dive spot. It was fewer than 100 yards (90 m) away from where we had dove before, so we were skeptical.

Toward the end of the dive we managed to find one wolffish, hidden in a hole under a huge rock. The only visible feature was its face, the mouth rhythmically opening and shutting, forcing water through its gills. Although this was just breathing to the fish, it looked to us like a vampire preparing to go for the jugular. We positioned our video camera and filmed the creature. The lights brought a slightly smaller wolffish to the mouth of the hole. Although the hole seemed too small for the pair of fish, it had obviously been carefully and intentionally cleared of debris inside to make room. Each fish tried to push the other aside to get a glimpse of the funny looking, bubble-blowing divers. They appeared especially curious, and we were delighted to have met this pair of fierce, but cautious, ocean inhabitants.

Later dives revealed several other holes that were homes to wolffishes. We always found a pair in

each hole. Even though at first the wolffish seemed to be alone, another would soon appear from within the deep recesses of the hole. In a week, we found some twelve pairs of wolffishes all living within 100 feet (30 m) of each other, but carefully hidden, in depths ranging from 50 to 120 feet (15 to 36.5 m). It seems likely that the Atlantic Wolffish mates for life, as its Pacific cousin is well known to do.

On some dives we experimented by trying to feed hot dogs to the wolffishes, who did not recognize them as food. Finally we offered them one of their natural foods: sea urchins. Although most of the fish would not eat from our offerings, we found one pair that would. We put the urchin on the end of a blunt-ended dive knife and put it into the opening of the hole. The wolffishes took the urchin in their powerful jaws, pulled it from the knife, and crushed it with the hard roof of their mouths. Then they swallowed the urchin, spines and all. We tried to tempt the wolffishes out of their holes with urchins, but they always refused to come out.

In an effort to photograph the fish outside their holes, we made a dive at night, when the animals search for food. We found empty dens, but no wolffishes. Where they went to feed was a mystery.

A year later, in need of some photographs for this book, I returned to New Brunswick. After

▲ *In its juvenile form, the Red hake (**Urophycis chuss**) sometimes burrows into the sand for safety.*

several unsuccessful dives, I found only three pairs of wolffishes, and none was interested in poking their heads out for a portrait. I had to photograph them with a telephoto lens, which puts a lot of water between the camera and the subject, detracting from the color saturation, contrast, and sharpness. The photographs that were obtained were adequate to show what the animals look like, but they were disappointing. On this trip, I did find the same pair of wolffishes that, the year before, had been so fond of eating urchins from the end of my dive knife. This time, I decided to try feeding them without the knife. I simply offered the urchin by holding it in two fingers. Like most fish, a wolffish opens its mouth and sucks the food in before biting down. In this way, the animal gently plucked the urchin from my hand and chomped it down. I felt no more threatened than when feeding my dog a bone (actually, a little less threatened. You should meet the dog). The stories of wolffish attacking people seem to come from fishers and spearfishing divers. It is likely that if you hook or spear a fish that has really big teeth, it will try to bite you. Although that trip was a photographic disappointment, seeing the pair of wolffishes from the year before and feeding them a few urchins really made my weekend. It was one of those magical moments that happens all too infrequently in the ocean.

◄ *Although most puffers are tropical fish, the North Atlantic is home to the Northern puffer (**Sphoeroides maculatus**) which grows to a sizable 10 inches.*

▲ *The most valuable of all tunas in the world is the Yellowfin Tuna (**Thunnus albacares**). It is frequently found near pods of dolphins, leading fishermen to capture and inadvertently kill dolphins while trying to get the tuna. This fish, photographed in the Gulf Stream off of North Carolina, is an extremely rare sight for underwater photographers.*

MARINE MAMMALS

◄ *Who can resist the charm of a Gray seal pup?*

No other group of marine animals thrills humans as large mammals do. Our close evolutionary kinship may explain people's interest, which may also be a response to these creature's impressive size. The plight of endangered marine mammals generates much worldwide attention, but they are difficult to study and remain poorly understood.

In general, a mammal is a vertebrate that bears live young, has hair, and is warm-blooded. There are, of course, a few exceptions, most notably the Australian Duck-Billed Platypus, a mammal that lays eggs. Mammals seem to have evolved from a small, four-legged, furry creature that lived in the undergrowth as the dinosaur age began to wanc. Dinosaurs were becoming extinct, while mammals prospered and evolved.

Marine mammals appear to have evolved from those same early ancestors and were, at one time, land animals. The most obvious indication

that these creatures evolved from land mammals is that they all have lungs and breathe air. The connection to land animals is apparent in seals, which have retained the four limbs and tail so characteristic of land mammals and look so much like undersea dogs. The flippers of seals still have five finger bones, and many species of seals still have five nails on their foreflippers.

Whales lost their rear limbs in favor of a horizontally flattened fluke (the term for a whale's tail fin) but still retain front flippers with finger bones. To emphasize the connection between whales and land mammals, let me point out that whales share about 90 percent of their molecular sequences with hippopotamuses.

To make the sea their home, many physical and physiological changes occurred in marine mammals. For example, compared with land animals, marine mammals produce more blood as a percentage

*A Harbor Seal (**Phoca vitulina**)* ▶
playing in shallow water.

of body weight and more hemoglobin in their blood to store more oxygen. Thus, the animals can remain submerged for extended periods without breathing. Whales and seals do not always hold their breaths as a human does. Instead, they store oxygen in their blood and muscles while on the surface, exhale completely, and then submerge. Using this technique, elephant seals can dive to depths greater than 2,000 feet (610 m) and stay submerged at least forty minutes. In one study, an elephant seal remained underwater without breathing for two hours.

Many marine mammals are very large. Baleen whales are the largest mammals in the world. Because they need to strain plankton from the water to survive, they must feed in the plankton-rich waters of the temperate and polar seas (see Introduction). Since they are warm-blooded and must maintain their body temperature in cooler waters, these whales must be physically large. Large animals retain body heat because they have more volume in proportion to their surface area than do small animals. This point is easily illustrated with a simple example. A cube with a measurement of 1 inch on a side has a surface area of 6 square inches and a volume of 1 cubic inch (a ratio of 6:1). A cube with a measurement of 100 inches on a side has a ratio of 6:100. The larger an animal is, the easier it is to hold a steady core temperature because heat loss through the skin is comparatively smaller. The disadvantage for the large animal is that it needs more food to survive.

Every marine mammal that lives predominantly in polar regions (including seals and whales)

is larger than its relative in warmer regions. For example, arctic seals are large, while tropical seals (like the Hawaiian Monk seal) are smaller.

Marine mammals also have a fat layer, which serves as insulation, aids in buoyancy control, and even provides an emergency energy reserve. Unfortunately, blubber has almost been the downfall of marine mammals, since its value has led to the mass hunting of whales and seals. Before the widespread use of petroleum oil products, whale blubber provided oil for lighting and lubrication and was used to make other products, such as perfume. Under United States law, the Marine Mammal Protection Act of 1972 now protects all marine mammals in U.S. waters from being hunted or harassed.

 ## THE SEALS

While whales and dolphins have adopted an entirely marine life, seals maintain a tie to the land. They come ashore to rest and to bask in the sun, and pups are born and nursed onshore. But for those exceptions, everything about the seal is geared toward its ability to survive in the water. Though awkward on land, in the water a seal is as graceful as an eagle soaring through the sky. Some seals make long migrations in the water, and some species of seals actually sleep underwater, but usually not for long.

Seals have a layer of blubber under the skin, and hair on their bodies. The hair's many oil glands increase the seals' insulation. During a dive, the

This is something rarely observed by humans: a ► Harbor seal sleeping underwater. Although it is known that seals sometimes nap for a few minutes underwater, it is rare for one to be in such a deep sleep that a diver can sneak up and take a picture. Only seconds after this photo was taken, the seal awoke in great surprise and swam away.

MARINE MAMMALS

circulation to most of the body except the brain and vital internal organs is severely restricted. The heartbeat slows to a rate as low as one-tenth the normal surface rate, and the body temperature is reduced to conserve oxygen. These adjustments allow seals to stay submerged as long as psssible. While submerged, most seals can swim at a rate of about 8 miles per hour. Seals see best underwater, having somewhat fuzzy vision on land. Even with their excellent underwater sight, however, vision is not essential for their survival. Many blind seals have been reported in perfect health, which suggests that seals use echolocation (using reflected sound waves to detect objects) to find food. We do know that seals produce such diverse (yet species-specific) sounds as clicking and groaning, which could be used for echolocation.

Seals possess acute hearing both in the water and on land. They can even close their ear openings and nostrils during a dive to keep water out. The seal's olfactory sense (smell) operates in air or water.

Seals have whiskers called Vibrissae, which, like those of a dog or a cat, are sensitive to vibration and touch. These whiskers are thought to play a role in capturing food.

All seals fall into the order Pinnipedia, which means 'wing-footed,' a reference to feet that have evolved into flippers. There are three families of seals: Eared (or Fur) seals (family Otariidae), the walrus (family Odobenidae, with only one species in the world), and True seals (family Phocidae).

The difference between Fur seals and True seals is in their body form and the way it is used in swimming and walking. Eared seals have muscular front flippers that can support the weight of the seal's upper body and lift its head off the ground. These front flippers provide swimming power, while the back flippers steer, like rudders. Eared seals can also rotate their rear flippers to point forward, which helps while walking on land.

True seals cannot lift their bodies or rotate their rear flippers forward and must move on land with an awkward crawl. In the water, however, True seals are extremely quick and use their rear flippers for locomotion rather than their front flippers. Sea Lions are an example of Fur seals, while the Harbor seal is a True seal. Strangely, the Atlantic is the only ocean that has no Fur seals, and they have never inhabited this part of the world.

SEALS IN THE NORTH ATLANTIC

The North Atlantic contains several different species of True seals, including Bearded seals, Gray seals, Harbor seals, Harp seals, Hooded seals, and Ringed seals. The Bearded, Harp, Hooded, and Ringed seals are all animals of the circumpolar Arctic; they are rarely found south of Labrador. These animals generally "haul out" onto pack ice (large chunks of floating ice) and tend to live near ice floes.

There have a been a few isolated reports of walrus being seen in the North Atlantic. Usually these reports come from extremely far north, such as Newfoundland, and rarely, from Nova Scotia. Normally these huge animals are found only in the Arctic.

THE HARBOR SEAL

Harbor seals (*Phoca vitulina*) are commonly found in the nearshore waters of northern oceans and adjoining bays above around 30° latitude. In the western North Atlantic, they are found from the Arctic (southern Greenland, Baffin Bay) to Maine, Massachusetts, New Hampshire, and occasionally as far south as Long Island. They will occasionally head a little more southerly in the winter, ranging from Maine to sometimes even the Carolinas.

Harbor seals haul out and bask on rocks and sand bars at low tide and forage for food at high tide. They eat fish and invertebrates, such as squid, herring, alewife, flounder, and hake. The Harbor seal is a superb swimmer and is able to catch fish easily. This fact has concerned anglers for years. Some fishers consider seals to be pests because they compete with them for fish. During the late 1800s, Maine and Massachusetts offered a $1.00 bounty on Harbor seals in order to increase fish densities. By the early 1900s, Maine's Harbor seals were nearly gone, with no noticeable effect on the fishing industry. The bounty was lifted in 1905 in Maine and, remarkably, the seals made a good comeback. Massachusetts maintained the bounty system until 1962, which may explain why seals have not yet returned to breed on Cape Cod. Harbor seals were last known to breed there during the early 1900s, and their current breeding grounds extend from the Arctic to New Hampshire.

Although it is almost impossible to approach adult ▶
Harbor seals, sometimes a pup will not feel
threatened by a human's slow, quiet approach.

MARINE MAMMALS

Mating occurs in water from early May to August. The males (bulls) will attempt to mate with many cows. Gestation takes a year, and pups are born from late April to mid-June. Each cow usually produces only one pup. The cow nurses the pup for about thirty days both in and out of the water. The pups can (and do) swim shortly after birth. The cow recognizes her pup both by its smell and by its call. When the cow is done nursing, she ceases giving any attention or protection to her pup and goes off to mate again. Once weaned, the pup (now called a yearling) is left to feed itself and must live on relatively slow-moving invertebrates until it becomes a sufficiently fast swimmer and skilled hunter to catch fish.

Pups born too late in the spring sometimes die because the mother leaves to mate again before the pup is completely weaned. Unable to hunt on its own, the pup starves. The first-year mortality rate of Harbor seals is about 30 percent, attributable to storms, abandonment, disease, parasites, and predation by sharks. Fortunately, mortality in later years is only about 13 percent. Harbor seals (in captivity) have lived to about thirty-five years old.

When fully grown, adult males average 5 feet (1.5 m) in length and weigh 200 pounds (90 kg) – $5^{1}/_{2}$ feet, 250 pounds (1.75 m, 113 kg) is their maximum size. Adult females average $4^{2}/_{3}$ feet, 156 pounds (1.25 m, 70.5 kg), $5^{1}/_{2}$ feet, 200 pounds is their maximum size. Males mature at four to six years, and females mature at three to four years.

Harbor seals molt in July and August. Like all seals, they shed their old coats and grow new ones. During that period, their skin looks like peeling sunburn; their fur flakes off in patches and new fur grows in.

Although seals seem playful and indeed can be, they are also cautious; harbor seals are particularly wary. It is nearly impossible to approach them onshore.

THE GRAY SEAL

The other seal common in the North Atlantic is the Gray seal (*Halichoerus grypus*). In the past these magnificent creatures have been called Horseheads or Horsehead seals because their elongated head looks like the head of a horse. In addition to a slight difference in coloration, Gray seals grow to be much larger than Harbor seals. Mature males reach 8 feet (2.5 m) and 800 pounds (363 kg), while the females reach 7 feet (2 m) and 400 pounds (181 kg). To distinguish between immature Gray seals and adult Harbor seals, one can look at their nostrils, which in the Gray seal are shaped like a W and in the Harbor seal like a V.

Gray seals are less common than Harbor seals and generally prefer isolated islands, especially to establish breeding colonies. Although Gray seals can be found from Labrador to Massachusetts, they usually do not breed south of Nova Scotia.

Massacred during years of bounty hunting, their numbers became dangerously low. Although they have made a remarkable comeback, they continue to be protected as an endangered species.

Gray seal pups are born during cold months. At birth they are just under 3 feet long (<1 m), weigh 30 pounds (13.5 kg), and are covered in long white fur. Within three weeks pups shed their fur and take on adult coloration. Although pups can swim at birth, they usually do not do much swimming until their first molt. Like the Harbor seal, pups are weaned from their mothers' milk in only two to three weeks. Since pups are born in the winter, when the water is especially cold, pups need to grow a significant layer of blubber in those few weeks. That is possible because the Gray seal's milk is among the richest of all mammals (it is up to 60 percent fat). Suckling this milk for five to six minutes every few hours allows the pups to add three or four pounds (1.5-2 kg) a day. Pups reach 90-100 pounds (40-45 kg) by the time they are weaned and without growing any longer. During the nursing period the mother eats little, and all the fat in the milk is drawn from her blubber reserves.

When Gray seals dive, they slow their heart rate to conserve oxygen, a process known as bradycardia. One study found a Gray seal that slowed its heart rate from 115 beats a minute at the surface to only 2 beats a minute for a fourteen-minute dive. Studies also show that Gray seals prefer traveling underwater because it is more efficient than swimming on the surface. Gray seals can stay submerged for more than twenty-five minutes and easily reach depths of 300 feet (91.5 m). It is estimated that Gray seal males live to age thirty-five, and females to forty-five.

◄ *Gray seal pups are born in the middle of winter. They have long white fur to keep them warm while they are nursing. In just three weeks after birth, the pups will shed their fur, and go off on their own.*

While filming seals for an Oceanic Research Group project, I was "discovered" by a small herd of Grays near Bar Harbor in Maine. The Grays were mixed with a herd of Harbor seals in a protected cove. We intended to shoot underwater video of the Harbor seals, not knowing there were any Grays there. The result was quite an encounter. I'll let excerpts from my log tell the story:

Tuesday: After almost half an hour, we moved just 25 feet and and suddenly a large Gray seal appeared. It moved in toward us cautiously. Phil Kelley [a dive assistant] held out his hand, which the seal took into its mouth and began to chew on gently. I was flailing a bit to get into position for a good photo angle. Finally I got the shot lined up and moved closer to the seal. After a few minutes of playing with Phil, the seal noticed my bright blue fins. The seal just loved those fins. He [it was a male] kept pulling on them and smelling them. I passed the camera to Phil and let him film the event.

Shortly afterward, the same seal returned and began playing with my fins again. I did not even see him coming. Instead, he swam up behind me and just grabbed a fin. I thought it was Phil pulling on it until I turned around. I'm not sure who was more startled, me, because something had grabbed my foot, or the seal when I quickly faced him. This time the seal was even less wary. He tasted my dry-suit leg and found it unappealing, so he moved up my leg and put his large mouth around my knee. I decided to offer him my hand, which he took. These are huge, powerful wild animals that could literally tear a diver apart underwater, but they do not. They do not bite, or nip, or

 # MARINE MAMMALS

make any threatening gestures. We do not coax them with food, nor do we chase them (that would be impossible). They come to us out of curiosity, and for some reason, I feel that I can trust them. I never felt even slightly afraid of the seal. It made me feel good that we could be friends, so to speak, even after all the things that people have, in the past, done to the seals. It is hard to explain other than to say that the experience really was moving. And the seals are just so darned cute!

Throughout a two-week filming period, seals came near us on only two days. On other days we saw no seals at all underwater, even though we knew that they were close by. The seals can see and hear us long before we can see one of them. They are at home in the sea, masters of their realm, and finely attuned to their surroundings, while we are intruders – aliens journeying out of our world to make contact with another species. To the seals, we are merely curiosities. They approach us with inquisitiveness, but also with caution. Any rapid movement or loud noise is interpreted as a possible threat, and they retreat. Several times I accidentally banged my tank on a rock or on the sea bottom, and each time the seals would scatter.

In February 1994, Oceanic Research Group traveled to Canada to film Gray seals pupping and mating. It was a trying trip because the seals seek to give birth at extremely remote places, and this isolation, coupled with the winter weather made reaching the seals difficult. We waited several days for the weather to allow us to board the boat and head for the island with the seals, but every day

there were problems. The harbor was frozen due to below-zero temperatures, and the wind howled fiercely, instantly freezing the ocean spray onto everything in its path. On the last day, we decided to try to get to the seals, no matter what.

We loaded the boat in the protected harbor and managed to get out through paths broken in the ice by a large ice-breaking cargo ship. As soon as we were out of the protection of the bay, the seas grew. Our bow plunged into the ocean on the downside of each wave. The wind whipped our faces and gear. Ice began to form on the boat, and it was clear that the weight of accumulating ice would soon sink our boat. It was hopeless, and we turned back. We consoled ourselves with the knowledge that if seals were easy to film, everyone would do it. Seals breed in isolated places to be safe from humans.

In January of the following year, we tried again. The weather was poor but warmer. We could tolerate rough, but not freezing, water on the boat. The harbor was still open, and we were absolutely committed to filming the seals, but there was another complication. An aerial survey showed that the pups had been born two weeks earlier than usual. That meant beginning our trip two weeks earlier than planned, and that would only allow one day of filming. Read the excerpts from my log:

At the harbor we loaded up a 24-foot fishing boat with far more gear than it had been designed to hold and keep dry. It was windy and rough. We were all being splashed with icy sea water. To avoid the salty shower, Brian Skerry and I hunched under a tiny cloth cabin in the bow. Tom Krasuki filmed away in the spray.

Our island destination finally loomed ahead, and the beach was littered with fuzzy white shapes. We had finally found our seal pups. How would they react to the boat? Would the pups try to swim away? Would the mother seals be protective of their pups? Would the bulls hold their ground? These were questions we wanted to answer.

The captain landed the boat downshore from the seals, and we quickly unloaded our gear. I was shooting stills, as was Brian, while Tom was shooting video for a new O.R.G. film. I gathered up my gear and headed into the icy grass, stomping my way back toward the seals. When I was within 100 yards I got down on my hands and knees and gathered up a camera. I stuffed an assortment of lenses and a few extra rolls of film into my heavy coat, grabbed a small tripod and started crawling toward the beach.

I crept closer, until my quarry lay only 10 feet ahead: a small bundle of white fur snoozing away peacefully on its side, directly ahead. I lined up the camera and took a shot. Nothing happened. My camera had frozen. Furious, I took off my gloves and examined the camera. Nothing seemed wrong. I put the camera inside my jacket and waited. My hands felt frozen, and I managed to stuff them back into my gloves.

There, directly in front of me was the picture I had in my mind's eye: a cute baby seal asleep and without a care in the world, and my camera was jammed. You can probably imagine my frustration.

After what seemed like an hour, I took the camera out of my jacket and tried it again. Nothing. I considered crawling all the way back to my pile of gear and grabbing another camera. As a last-ditch effort, I

Gray seals mate on land or in the surf during the winter just after the pups are born. ▶

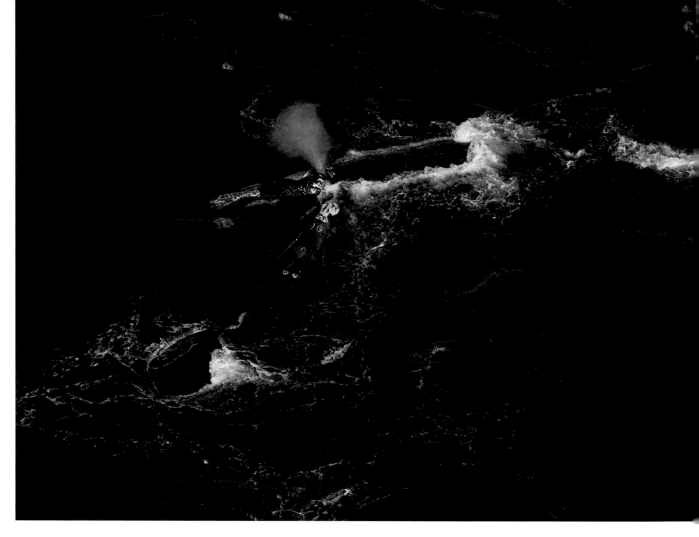

wacked the thing on its side and with a "pop," it unjammed. (I call this field camera repair.)

Just as I hit the camera, the drowsy, teary-eyed baby seal awoke and looked at me. I had no idea what to expect. I was only 10 feet from this wild animal, which had never before, in its two-week-long lifetime, set eyes upon a human being. What would it do?

The seal yawned and rolled over, falling right back to sleep, but now facing the wrong way. I crawled around and started taking pictures. I continued to photograph other pups, in various stages of molt, for a few hours. Once a baby seal woke up surprised to find me only a few feet from her. She growled, I backed up, and then she fell right back to sleep. Some pups did not mind me at all, and others were happy as long as I stayed a few feet away. All of them got used to me in a few minutes and let me photograph freely. It is easy to see how seal fur traders can so easily kill seal pups. You can walk right up to them. Perhaps these seals will remember me as a friend, not as a predator.

Eventually, the pups began to wake up and cry. They howled and moaned sorrowfully for their mothers. They were getting hungry, and I realized that I was keeping them from being fed. The mother seals were swimming just off the beach, keeping watch, but afraid to come near. Most of the bulls had moved down the beach or across the water to a ledge poking out of the ocean only 100 yards out.

I began backing away from the beach and finally settled into the bone-chilling sand, about 75 yards away. The cows began to come out of the water. They glanced around for their pups, trying to distinguish between all the different calls. Then they looked

▲ *As seen from an airplane, two 50 foot long Northern Right whales prepare to mate.*

at me, trying to decide whether I was far enough away, or whether I meant them harm. Finally, one dedicated mom boldly romped up the beach and sniffed at a pup. It was obvious that the pup did not really care whether it was his mom or not, as long as he got lunch. The cow was not about to feed someone else's pup, however, and she moved on to the next pup and finally found her own. Rolling over on her side, she allowed her pup to nurse for several minutes, and the contented pup went right back to sleep.

When people see my photographs of these pups, they often comment on how adorable they are and ask me whether I petted them. The answer is no. The pups might be upset, and the relationship between the pup and its mother might be disturbed. If mothers recognize their pups primarily by smell (as most scientists believe), then touching a pup might change its smell enough to confuse the mother. They are just as cute from a few feet away, and a hands-off policy is best.

THE WHALES

Calling the whales a big family is an understatement. Several rank as the largest creatures on Earth. In fact, the Bluc whale is the largest known creature ever to inhabit the Earth. Not even dinosaurs reached the size of the mighty Blue whale (*Balaenoptera musculus*), stretching 100 feet (30 m) long and tipping the scales at up to 200 tons (it weighs more than 2,500 people). Yet, not all whales are large. The Harbor porpoise (*Phocoena phocoena*) reaches only 6 feet (<2 m) in length.

Their size does not protect them, for whales have been hunted to the brink of extinction. Unfortunately, hunting has also provided the only opportunity to study these creatures, at least in the past. The Right whale (see below) takes its name from whaling tradition: it was the "right" whale to hunt because it swam slowly and floated when dead. Only recently have scientists begun to study whales in order to learn more about their world and their lives.

Some of the discoveries made about whales are fascinating. For example, the Humpback whale carefully repeats its songs many times in distinct phrases and segments, much like our music. We do not know why they sing. It could be a mating ritual, a form of communication, or even an artful expression. The possibilities are numerous. After all, many of the great whales have much larger brains than humans and perhaps even more complex ones as well.

Many whales and porpoises use echolocation to find prey. Experiments have shown that these animals can "see" for many miles underwater using echolocation. In addition, the echolocation is so sensitive that a Common dolphin (*Delphinus delphis*) can differentiate between two metal objects that differ by only 1/8 of an inch (3 mm) in thickness. Common dolphins can also produce sound underwater with their echolocation system loud enough to stun a fish. It has been suggested that in the days before loud underwater sounds generated by ships and motorboats, the great whales may have communicated with each other across the entire ocean, or even around the world, by producing loud tones that carried through the water.

Actually, we know little about whale behavior. Why, for example, do whales breach (jump out of the water and splash back in)? Biologists can speculate, but we simply do not know with certainty.

TAXONOMY

All whales (including porpoises and dolphins) belong to the order Cetacea. This order is further divided into two suborders, the Mysticeti and Odontoceti. The Mysticetes (more commonly called baleen whales) feed on plankton, small fish, and other small creatures by straining them from the water using baleen plates. Baleen plates are made of keratin, the same tough material of which human fingernails are made. The Odontocetes have teeth and feed on many different types of animals. This suborder includes dolphins, porpoises, and orcas ("killer" whales).

Many whales migrate long distances between feeding and birthing grounds; others seem

▼ *Reaching nine feet in length, the Atlantic White Sided dolphin* (**Lagenorhynchus acutus**) *are sometimes found in very large pods, but do not frequently ride bow waves. Here, two females and their calves swim with their pod in the early morning.*

to wander the oceans randomly. Whales do not always restrict themselves to the relatively shallow waters of the continental shelf (although that is where much of their food is found). Whales, like people, can be unpredictable. For example, orcas, generally known to be cold-water whales, occasionally appear in the warm Caribbean waters.

THE MYSTICETES OF THE NORTH ATLANTIC

The suborder Mysticeti contains three families: the Rorquals (family Balaenopteridae), the Right whales (family Balaenidae), and the Gray whales (family Eschrichtidae). Representatives of the Rorquals and Right whales are found in the North Atlantic, but there are no Grays. This has not always been the case, however.

The Atlantic Gray whale was hunted to extinction before the species was ever studied or scientifically described. We have no photographs of these animals, only a few drawings and written descriptions from whaling captains. The Atlantic Gray is lost forever, and only its Pacific cousin remains.

THE RORQUALS OF THE NORTH ATLANTIC

Among the Rorquals found in the North Atlantic are the Blue whales, Fin whales, Humpback whales, Minke whales, and Sei whales. The Rorquals are long and rather skinny. Their throats have external grooves, which allow the throat to expand when the whale takes an enormous gulp of water while straining for plankton or fish.

The Humpback whale (*Megaptera novaeangliae*) is probably the most notable Rorqual in the North Atlantic. It grows to more than 50 feet ($15^1/_2$ m) long but has no hump. Its name comes from the way it arches in the middle as it dives, then throws its fluke from the water. The genus Megaptera literally means 'big-winged,' and refers to the size of the Humpback's flippers, which are the largest appendages in the animal world and equal one-third of the whale's body length. The Humpback is well known for singing long, elaborate songs, which it repeats over and over. Scientists are not sure of the songs' meaning, but they are most likely a means of communication.

Atlantic Humpbacks migrate annually from their winter calving grounds in the Caribbean to their summer feeding grounds in the North Atlantic. Although Humpbacks give birth in the warm waters of the Caribbean, throughout the winter they do not eat, because there is not enough food in the warm tropical waters to feed them. They migrate back to the cooler waters of the North Atlantic in the spring when the calves have added enough blubber to keep them warm. Humpbacks, like most whales, have only one calf at a time, and the calf is born fluke first so that it will not drown during birth.

◄ *Swimming with Northern Right whales while they attempted to mate made my heart pound with excitement. Even though they are so large, the whales still knew I was there tried to keep from injuring me with their rough foreplay.*

The Humpback is a favorite of whale watchers in the North Atlantic because it frequently performs fantastic breaches and spyhopping (poking its head out of the water). Humpbacks also approach whale-watching boats for close inspection. Some biologists think that the sound of boat engines makes the whales curious. Humpbacks are easy to view since they tend to prefer waters relatively close to shore to the continental shelf. Although this is good for whale watchers, it was fatal for Humpbacks during the whaling era, when they were easily found by hunters. As a result, the Humpback population was reduced almost to extinction in both the Atlantic and Pacific. It has been protected worldwide since 1966, but its world-wide population today is estimated at only twelve thousand, which is less than one-tenth its prewhaling numbers.

The Minke whale (*Balaenoptera acutorostrata*) grows to 33 feet (10 m), which may still seem large but which makes this animal the smallest Mysticeti in North American waters. Minkes occupy the same continental shelf waters as do Humpbacks and are found from the Arctic to the tropics.

The closely related Fin whale (*Balaenoptera physalus*), which grows to 80 feet (24 m) long, has the distinction of being the world's second-largest whale after the Blue whale. These huge animals are fairly common (by whale standards) in the North Atlantic. In 1993, I visited the Mount Desert Rock Whale and Seabird Research Station, operated by the College of the Atlantic. The scientists there were actively tracking (through photographs) over one

hundred Fin whales. These whales are, like Humpbacks, known to breach on occasion, and what a splash they make!

THE RIGHT WHALES OF THE NORTH ATLANTIC

In the North Atlantic, the Right whale and the Bowhead whale both belong to the family of Right whales. These whales lack the ventral grooves found on Rorquals, which allow expansion of the throat, and they also lack a dorsal fin. The head is large – as much as 40 percent of the total length of the animal – and rounded, giving the whale a somewhat curvaceous appearance.

The Bowhead whale (*Balaena mysticetus*) grows to 65 feet (18 m) and is found only in the northernmost Atlantic, around Greenland and Baffin Bay, rarely straying from drifting pack ice. Relatively little is known about this species, as its population, an estimated eight thousand individuals remaining in the entire world, is extremely endangered.

The Northern Right whale (*Eubalaena glacialis*) grows to about 55 feet (16.5 m) long and is found in the Atlantic from Iceland to Florida. Being the "right" species to hunt, this whale is close to extinction. Before the whaling era, an estimated 50,000 Right whales roamed the oceans of the world. That total included both the Northern Right whale and the Southern Right whale (*Eubalaena australis*), found in the South Atlantic and Pacific. It is estimated that there are now only about 3,000

▲ *The Northern Right whale (**Eubalaena glacialis**) is the most endangered whale in the world. Its population is just 350 in the North Atlantic and probably not more than about 500 worldwide.*

MARINE MAMMALS

Southern Right whales, and their comeback has been excruciatingly slow, even by whale standards. Even more seriously decimated, the population of Northern Right whales in the Atlantic is estimated at only 350 individuals, making it the most endangered whale in the world.

In 1980, Scott Kraus, a researcher from the New England Aquarium, discovered the Northern Right whale's summer feeding and mating grounds in the Bay of Fundy. The location of its wintering grounds was still a mystery, but whaling records from the 1800s revealed that whales had been killed off the coast of Georgia. Aerial surveys confirmed the Northern Right's presence there today, although in greatly reduced numbers.

The Northern Right whale is perhaps the slowest swimmer of all whales and therefore is sometimes hit by ships. During the twentieth century, ship collisions became one of the more common causes of death among Right whales. In the summer of 1994, for example, a Northern Right whale was killed after being struck by a ferry in the Bay of Fundy. Every Northern Right whale death is a terrible blow to the species. Because the small population is growing so slowly, the Northern Right whale is in grave danger of extinction. Every entanglement in fishing gear, every accidental collision, and every death by pollution brings the gentle Northern Right whale closer to the brink of doom.

DNA testing of Northern Right whale tissue has suggested that at some time within the past five hundred years the population of breeding females in the North Atlantic may have been reduced to just three individuals. The entire population alive today is descended from one of those three whales and their mates.

In 1995, O.R.G. divers traveled to the Bay of Fundy to photograph the Northern Right whale and discovered a beautiful creature worthy of our protection and care. Flying over the bay in a rented Cessna, we photographed whales mating. Passionate expression is an ordeal for this species. When a female is ready to mate, she signals the males who have gathered in the vicinity. The female holds her breath and rolls over onto her back, which lifts her genital slit out of the water, where the males cannot reach it. Finally, in need of air, the female rolls over to breathe. At this point, all the males rush in and try to mate, unlike male Humpbacks, who fight each other fiercely for the right to mate. The successful Northern Right suitors are the ones with the most well-developed genitalia (they have an 11-foot-long [3.5 m] penis, and testicles which may weigh a ton and produce copious amounts of sperm). This is natural selection at work, because only the whales with the largest amounts of sperm and the best penis placement will impregnate the female. If this competition sounds violent, it is. Three or four 50-ton animals thrashing around in the water can make quite a splash. You can imagine the look on the pilot's face when we told him that we planned to dive with those whales in their mating frenzy.

With permits from the Canadian government, we set out the next day with our gear in a

▲ *Although they prefer arctic waters, Beluga whales (**Delphinapterus leucas**) are not unknown in the North Atlantic and Gulf of Maine. Their white coloration helps them blend into the sea ice of their normal arctic habitat. In a chance encounter, this 11-foot-long female approached the author while diving in a remote bay off Nova Scotia.*

MARINE MAMMALS

small boat. After riding several hours into the bay, we found four Right whales who were mating. Stopping the boat about 100 yards away, we slid into the water and began swimming toward the whales. Visibility was low (as little as 10 feet) because the water was green with phytoplankton, food for the small zooplankton that whales eat. We swam closer, until we were practically on top of the whales, but we could see almost nothing underwater. Since we were concerned that the sound and bubbles of scuba gear might frighten the whales, we used snorkels instead.

To describe my first encounter with a Right whale underwater is almost impossible. I put my mask in the water as a whale swam past, and he looked at me carefully. I dove down to swim alongside, and we looked into each other's eyes. Swimming much faster, he passed me and carefully avoided striking me with his fluke. There is no doubt that if a whale played too roughly with me, I could be seriously injured or killed, but I felt I could trust these whales. They seemed as curious as I was, except when the female into mating position. Then I found myself a little too close and was bumped by an enormous pectoral fin and shoved sideways. Although it was frightening at first, I realized that the whales knew I was there and were actually trying to avoid me. The grace and maneuverability of a 50-foot-long whale is remarkable.

It is easy to forget just how closely attuned whales are to their surroundings. I was reminded of that fact by an amazing encounter. While treading water, I saw a Right whale heading my way. So I held my breath and swam down 10 feet to meet the whale. He came up and stopped, apparently giving me a good looking over. I raised my camera and clicked the shutter. The whale flinched. I actually startled a whale with the click of a shutter. He forgave me and even grew accustomed to the camera as I fired off a few more shots.

When Tom Krasuki and I got back into the boat, the captain could not believe what he had just seen. He was astonished that the whales were so gentle with us and that they were curious. Of course, not all the whales we encountered that day were curious. Most, in fact, were indifferent and wanted only to feed or mate. When we got into the water with those whales, they just swam away, leaving us alone in the sea. Although it is possible to harass a whale with a boat, it is impossible to harass a whale with snorkel or scuba gear. Even the slowest whales can swim away any time they want and leave us in their wake. Encounters with whales underwater happen only on their terms, and that is the way it should be.

THE ODONTOCETES OF THE NORTH ATLANTIC

The suborder Odontoceti contains six families of toothed whales: Beaked whales (family Ziphiidae), Narwhal and White whales (family Monodontidae), Ocean dolphins (family Delphinidae), porpoises (family Phocoenidae), River dolphins (family Platanispidae), and Sperm whales (family Physeteridae).

Beaked whales are a somewhat mysterious family about which little is known. There are at least five species in the North Atlantic, all of which are medium in length (approximately 15-30 feet, or 4.5-9 m).

The Narwhal and White whales are animals of the far north, living among ice floes, which they resemble in color. They are still hunted legally for food by Eskimos, but their survival is not threatened. White whales like the Beluga (*Delphinapterus leucas*) are gregarious and sometimes gather in pods of up to a thousand individuals.

The Ocean dolphin family has many well-known (and some not so well-known) members in the North Atlantic, including (but not limited to) the Bottlenosed dolphin, the Common dolphin, the Orca, and the Pilot whale. Pilot whales and dolphins are common throughout the North Atlantic. Many lucky boaters have been both captive audience and host to a pod of dolphins riding the bow wave of their boat. Dolphins may do this for enjoyment or because they enjoy the free ride. Dolphins will sometimes play with divers for a few minutes and then swim away when they get bored.

There have been stories for thousands of years about dolphins saving the lives of sailors lost at sea. Although these stories may be exaggerated, dolphins clearly are not hostile toward people. Dolphins have shown that they are smart and can

When mating, male Northern Right whales compete for ▶
the best positions near the female, sometimes violently
splashing the water with their enormous flukes.

learn complex commands. The United States Navy has trained dolphins to retrieve objects lost on the ocean floor and to carry objects back and forth to divers working in the ocean.

There is a particularly unusual Bottlenosed dolphin (*Tursiops truncatus*) named JoJo who lives in the Turks and Caicos Islands in the Caribbean Sea. He is not a member of a pod. Dolphins are "social" animals and like to be with other dolphins. A dolphin without a pod is a true rarity in nature. JoJo prefers human company. He stays around the local beaches and plays with swimmers. One of his favorite tricks is to swim between people's legs when they least expect it. Sometimes his playing is a bit rough, but scientists believe that JoJo is, in fact, playing. It is important to remember that JoJo is an unusual dolphin, however, and is not representative of most dolphins. I visited the Turks and Caicos in 1993 and met JoJo, who was still playing with the swimmers at the beach.

Many people have seen dolphins (or other small cetaceans) perform in captivity, and opinions about this are divided. Opponents believe that captivity for dolphins is cruel, especially for an animal that possesses echolocation, allowing it to "see" for miles and miles underwater. Putting such an animal in a tank, no matter how big, is like putting a person into a phone booth painted black: it represents a complete closing in of the senses. Few dolphins taken from the wild and placed into tanks survive. A social animal like the dolphin needs the company of other dolphins in the open ocean. The

only dolphins that have done well in captivity are those raised in captivity from birth. Proponents believe that keeping dolphins in captivity is the best way to prompt human concern about the plight of dolphins. By seeing real dolphins, people may appreciate them more.

There is a tribe of people on Africa's Atlantic coast, called the Imragen, who count on the assistance of dolphins to catch mullet. Each year between December and February, large schools of mullet migrate through the region where the Imragen live. Dolphins, apparently following the

schools of fish for food, are never far away. When a school of fish passes by the beach, the Imragen pound the water with sticks to attract the dolphins. The dolphins then steer the school of fish toward shore, where the fish are caught in nets. Are the dolphins intentionally helping the Imragen people or are they just trying to get a meal themselves? The tribe believes that the dolphins are helping them, and the dolphin is sacred in their culture.

The Orca (*Orcinus orca*), sometimes called the Killer whale, can be found in the North Atlantic, although it is more common in the North

MARINE MAMMALS

Pacific. A small population exists in the Atlantic from the ice pack of the Artic south to Nova Scotia and can sometimes be found as far south as the Caribbean. Orcas prefer to eat fish, seals, and other whales (particularly their blubber and tongues). One of the most remarkable facts about orcas is how they hunt. Orcas travel and hunt in family groups and cleverly work together to capture food. When a pod of orcas approaches a pod of whales, they attempt to isolate one individual from the pod and cooperate well in completing this task. Orcas have also been known to swim out of the water and up onto the beach to catch seals splashing in the surf, wiggling back into the sea with their meals. Orcas reach 31 feet (9.5 m) in length.

Porpoises are the smallest family of cetaceans in terms of body size. Only the Harbor porpoise (*Phocoena phocoena*) is common in the North Atlantic. This small, toothed whale reaches about 6 feet (<2 m) in length and prefers cold coastal waters from Greenland to (occasionally) as far south as North Carolina. They are common throughout New England. Unlike dolphins, these animals do not ride bow waves and generally avoid boats and people. As a result, they have been not well studied.

River dolphins are freshwater cetaceans found in the Amazon and China, among other places, but not in North America.

Most people have at least heard about the Sperm whale (*Physeter catadon*) because of its role in the classic novel *Moby Dick*. Beyond its fictional role, however, the Sperm whale is a remarkable creature. It has the largest brain of any animal on Earth, leading many scientists to speculate that it may be intelligent. We do know that Sperm whales can remain submerged for more than an hour and are believed to dive to depths greater than 7,000 feet (2,134 m) to hunt for their favorite food, giant deep-sea squid. To put that depth in perspective, 7,000 feet of seawater produces a pressure of more than 200 atmospheres (or 3,000 psi), which is out of reach for all but the deepest-diving of research submarines. The future of the Sperm whale is promising. More than 1 million of these magnificent animals remain in the wild. Their numbers, now at about one-half prewhaling totals, are growing significantly.

MARINE MAMMALS AND THE LAW

Fortunately, all marine mammals enjoy full protection under United States law. However, whales do not always stay in safe U.S. waters; many are long-distance migrators. And because several countries do not accept the International Whaling Commission (IWC) ban on whale hunting, the oceans continue to be a dangerous place for whales. Norway and Japan still hunt whales, and in Japan, dolphins and porpoises are rounded up by the thousands, driven toward shore and bludgeoned to death with clubs. Canada continues to allow the wholesale slaughter of Harp and Gray seal pups.

Recently, environmental organizations have brought the issue to light with a publicity campaign showing cute little pups being bludgeoned to death. Although Harp seals are no longer endangered, Gray seals continue to be at risk. Seal pup fur is no longer used for any product, but the destruction will not cease as long as fishers believe that the small amount of fish consumed by seals reduces their catch. The killing of seals is needless and should stop now.

If we are ever to learn about marine mammals and the things they may teach us about the sea and ourselves, we must protect them from human predation around the globe. The reproduction cycle of many marine mammals, especially whales, is slow, which makes any comeback from hunting a long process. We must not allow the same fate of the extinct Atlantic Gray whale and Caribbean Monk seal to befall other great whales and marine mammals.

EPILOGUE

When the Pilgrims came to New England they discovered that the North Atlantic contained vast quantities of fish. In letters back to England, they claimed that a person didn't even need a hook to catch them. In fact, they said that with a simple bucket one could easily scoop fish out of the water. After living near overfished European waters, the Pilgrims were unaccustomed to densities of fish like those in New England's ocean cornucopia. Not only was the North Atlantic an incredibly fertile area for fish to grow, no one was fishing the stocks.

Because it was so plentiful, Codfish became a favorite food of the settlers. It was abundant, and it was tasty. The Pilgrims were soon salting it to preserve it and exporting it to England. Salted cod was one of the few commodities the settlers could trade for goods from England that they needed. Soon, however, word got out about New England's bounty and English boats began fishing for it. Eventually, Spain, France, Portugal, Germany and other countries sent ships as well.

Contrary to popular beliefs at the time, the sea was not so large that limitless supplies of fish to be taken from it. In fact, 95 percent of the fish live in only 4 percent of the ocean. The waters of the continental shelves hold most of the fish we consume.

The fish harvests began to wane only a hundred years after the settlers first arrived, but no one thought that the fish were running out. Instead, it was thought that they were moving somewhere else to avoid the fishermen. So, the fishing fleets spread out. They went further north and further south. In many cases, new "virgin" fish populations were found which held out for a while. The fishermen, however, believed that they had found the same old schools of fish in new places.

As fish stocks continued to dwindle, fishermen employed more ingenious tools to catch fish. First there was the hook and line. Then came the line with many hooks, then came the net, which itself was specialized depending on which type of fish was sought. Some were pulled behind a boat to scoop up pelagic fish, some were dragged along the bottom to scoop up bottom dwelling fish, and some were left to drift and were checked periodically. Then came electronic fish finders based on SONAR technology. Today, we have so much technology to find fish, that we are literally taking every fish in the ocean.

Vito Giacalone, a fish population specialist at the Massachusetts Marine Fisheries Department sums it up like this,

"Over the years, technology has been refined tremendously. It's gotten to the point today where fisherman can literally catch every fish that's out in the ocean. The poor fish don't have a chance!"

The Codfish population has perhaps suffered the most of all. In 1992, Labrador and Newfoundland (whose economies are almost completely based upon fishing, and particularly Cod fishing) passed a complete moratorium on Cod fishing. It was met with resistance, anger, and even violence by the fishermen. They believed that if they fished more, they could find fish to make a living and feed their families. They felt that a moratorium was a personal attack on their way of life. But the Codfish is virtually gone from the area, fished into oblivion. Nobody had the forethought ten years ago to step back and say "Hey, maybe we should let up a little on the fish." Nobody dared say that to the fishermen, until it was too late.

Fishing has been lucrative in the United States. In good years fishermen make a lot of money. But not all years are good, and good years are becoming less common. The problem with fishing is that the laws of supply and demand tend to be bad for the fish. For example, when the Codfish population gets low, the supply is limited, and the price goes up. The fishermen all try even harder to catch Codfish because it is fetching such a high price at market. This increases the pressure on the codfish fishery, and pushes the population further downhill, as was the case in Labrador and Newfoundland.

The price of tuna has skyrocketed because it is sought after in Japan for sushi. At almost any fish pier in the U.S., Japanese representatives are ready to pay cash for tuna as it comes off the boat. A single large Bluefin Tuna may earn a fisherman $10,000. By the time it gets to Japan, it may be worth $30,000 dollars. The Japanese people will pay the equivalent of $50.00 per ounce to eat it! With pressure like this, how can the ocean's fisheries possibly survive?

Unfortunately, the American fisheries are headed in the same direction. Mr. Giacalone puts it well when he says: "If nothing is done to control the overfishing which is going on today, there will not be a [fish] resource in the Atlantic Ocean. The resources of Cod and Haddock are depleted to such a low state, that there's not enough fish to supply the need. So, the projection is that there will not be enough fish to feed the people in the United States if something isn't done about it today."

Since the well-publicized Codfish moratorium in Labrador and Newfoundland, many American fishermen have begun to acknowledge that this legislation may be their best hope for maintaining the fishery for the future. I have met fishermen who now agree that some "reasonable" limits have to be set. But what is reasonable?

Many foreign countries are much more dependent on fish to survive than we are in the United States. Americans eats fish less than one night a week on average, while in Europe or Japan, it ranges from five to seven nights per week. The United States has an incredible beef and poultry industry. We have the land and experience to raise cattle efficiently and profitably. Those skills and infrastructure weren't developed overnight. It took years to learn how to do it well, and we are still learning. Many other countries do not have the space, determination, resources or skill to raise beef and poultry on a large scale. Instead, they fish. Why is this fundamentally different? Because fishing is nothing more than hunting, while raising cattle is farming.

Five thousand years ago, in ancient Mesopotamia, man learned to farm cattle as a source of food. It has only been in the past decade or so, however, that man has begun to farm the oceans as a source of food. We have been hunting wild animals from the seas for thousands of years without ever trying to put back what we take.

Aquaculture is the wave of the future for the fishing industry because it leaves natural fish populations intact, and allows people to farm fish. Unfortunately, it has proven more difficult than farming cattle. The Japanese were the first to undertake extensive aquaculture, but the United States is also making strides in this new technology. For example, the Atlantic Salmon is commercially extinct in the wild (there are several nature groups trying to get this fish protected by law as an endangered species). They turn up here and there, but nobody fishes for them anymore. Fishermen have almost completely wiped out the natural stocks, but you can still go to the market and find Atlantic Salmon. It comes from farms, mostly in northern Maine and eastern Canada. Other people are trying to farm mussels, lobsters, lumpfish, oysters, crabs, and other popular sea foods. If aquaculture becomes as successful as terrestrial farming, we can expect this new science to provide much of our sea food needs at little cost to Mother Nature and her natural fish stocks.

Unfortunately, overfishing isn't the only threat which our oceans face today. An estimated 14 billion pounds of garbage are dumped into the ocean every year. That's more than 1.5 million pounds per hour! In one beach clean-up along 150 miles of North Carolina coastline, 8,000 plastic garbage bags were collected in three hours. Isn't it ironic that we think of the oceans as a place to harvest food, yet we also use it as a garbage dump? Unfortunately, we are beginning to see the inevitable results of this type of behavior in the form of poisoned, inedible fish and polluted beaches.

Coastal areas are inundated by massive agricultural run-off rich in nutrients, causing red tides and algae blooms which kill marine life. Oil spills seem a constant presence in the news. We don't yet know how much long-term damage they do, although the images from Prince William Sound, Alaska remind many people of the short-term damage. Marine pollution is a serious threat to our oceans, and ultimately, to our world. People once thought that the ocean was so large that it could absorb an infinite amount of trash and sewage with no perceptible change. We now know how wrong that thinking is. Every beach in the world, no matter how remote, is littered with trash from distant lands. The time has come to change our ways. We have the knowledge and the technology to live harmoniously with nature and the environment, but will we use that knowledge?

The ocean has a magical effect upon me and on most people. It occupies more than 70 percent of our planet's surface area, yet we are only now beginning to understand it. Every ocean is important for the world: for us and for the environment. Every ocean creature has something to offer us in some way. We must all fight to preserve our oceans and to make sure that every country of the world treats the oceans with the respect they deserve. After all, we humans are but a single representation of the enormous diversity of life living on continental islands in the sea of our Earth, the water planet.

BIBLIOGRAPHY AND FURTHER READING:

The Audobon Society Field Guide to North American Fishes, Whales and Dolphins, Alfred A. Knopf, Inc., 1988

Barnes, Robert D., *Invertebrate Zoology*, Fifth Edition, Harcourt Brace Jovanovich College Publishers, 1987

Bavendam, Fred, *Beneath Cold Waters*, Down East Books, 1980

Bigelow, Henry B. and Schroeder, William C., *Fishes of the Gulf of Maine*, Museum of Comparative Zoology, Harvard University, 1964

Buchsbaum, Ralph, *Animals without Backbones*, The Chicago University Press, 1974

Buchsbaum, R. and Milne L., *The Lower Animals*, Doubleday & Co., Inc., 1966

Cousteau, J.Y., and Diole, P., *Dolphins*, Arrowood Press, 1987

Emson, R.H. and Wilkie, I.C. 1980. "Fission and Autotomy In Echinoderms," Oceanography and Marine Biology Annual Review, Vol. 18, pp 155-250

Gilbert, Perry W., *Sharks and Survival*, D.C. Heath and Company, 1975

Gillis, Anna Maria, 1991. "Why can't we balance the globe's carbon budget?" BioScience, Vol. 41, Num. 7, pp 442-447

Gormley, G., *Orcas of the Gulf*, Douglas & McIntyre, Ltd., 1990

Hoyt, Erich, *Seasons of the Whale*, Nimbus Publishing Ltd., 1990

Iverson, S. J., Bowen, W. D., Boness, D. J., and Oftedal, O. T. 1993. "The Effect of Maternal Size and Milk Energy Output on Pup Growth in Grey Seals (*Halichoerus grypus*)," Physiological Zoology, Vol. 66, Num. 1, pp. 61-88

Katona, S., Richardson, D., and Rough V., *Field Guide to the Whales, Porpoises and Seals of the Gulf of Maine and Eastern Canada*, Macmillian, 1983

King, Judith E., *Seals of the World*, Cornell University Press, 1983

Macdonald, D. ed, *Sea Mammals*, Torstar Books, Inc., New York, NY, 1984

Meinkoth, Norman A., *The Audobon Society Field Guide to Seashore Creatures*, Alfred A. Knopf, Inc., 1988

Nicklin, F., and Darling, J., *With The Whales*, NorthWord Press, Inc., 1990

Parsons, T., Masayuki, T., and Hargrave, B., *Biological Oceanographic Processes*, Pergamon Press, 1984

Pennisi, E., 1993. "Chitin Craze," Science News, Vol. 144, No. 5, pp 72-74

Quayle, Louise, *Dolphins and Porpoises*, Gallery Books, 1988

Smith, DeBoyd L., *A Guide to Marine Coastal Plankton and Marine Invertebrate Larvae*, Kendall/Hunt Publishing Co., 1977

Stevens, John D., *Sharks*, Facts On File Publications, 1987

Sumich, James L., *An Introduction to the Biology of Marine Life*, Wm. C. Brown Company Publishers, 1976

Thompson, D. and Fedak, M. A. 1993. "Cardiac Responses of Grey Seals During Diving at Sea," Journal of Experimental Biology, Vol 174, pp 139-154

Thurman, Harold V., *Introductory Oceanography*, Sixth Edition, Macmillan Publishing Company, 1988

INDEX

INDEX

Note: bold page number indicates photo.